All About Alice

Exponents, Scientific Notation, and Logarithms

Teacher's Guide

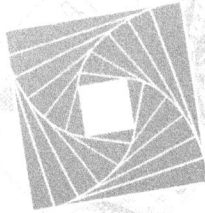

This material is based upon work supported by the National Science Foundation under award numbers ESI-9255262, ESI-0137805, and ESI-0627821. Any opinions, findings, and conclusions or recommendations expressed in this publication are those of the authors and do not necessarily reflect the views of the National Science Foundation.

Key Curriculum
1150 65th Street
Emeryville, California 94608
email: editorial@keypress.com
www.keycurriculum.com

First Edition Authors

Dan Fendel, Diane Resek, Lynne Alper, and Sherry Fraser

Contributors to the Second Edition

Sherry Fraser, Jean Klanica, Brian Lawler, Eric Robinson, Lew Romagnano, Rick Marks, Dan Brutlag, Alan Olds, Mike Bryant, Jeri P. Philbrick, Lori Green, Matt Bremer, Margaret DeArmond

Project Editors

Sharon Taylor

Consulting Editor

Mali Apple

Project Administrator

Juliana Tringali

Professional Reviewer

Rick Marks, Sonoma State University, CA

Calculator Materials Editor

Josephine Noah

Math Checker

Carrie Gongaware

Production Editor

Andrew Jones

Production Director

Christine Osborne

Executive Editor

Josephine Noah

Mathematics Product Manager

Timothy Pope

Publisher

Steven Rasmussen

Contents

Introduction

All About Alice Unit Overview

Intent

This unit uses Lewis Carroll's story *Alice's Adventures in Wonderland* as the context in which students define the exponential function, derive properties of exponents, and use exponents to solve problems.

Mathematics

Unlike most other IMP units, *All About Alice* has no central problem to solve. Instead, there is a general context to the unit, as in the Year 1 unit *The Overland Trail*.

In particular, the Alice story provides a metaphor for understanding **exponents.** When Alice eats an ounce of cake, her height is multiplied by a particular whole-number amount; when she drinks an ounce of beverage, her height is multiplied by a particular fractional amount. Using this metaphor, students reason about exponential growth and decay.

Students use several approaches to extend exponentiation beyond positive integers: a contextual situation, algebraic laws, graphs, and number patterns. They then apply principles of exponents to study **logarithms** and **scientific notation.**

The main concepts and skills students will encounter and practice during the course of this unit are summarized by category here.

Extending the Operation of Exponentiation

Defining the operation for an exponent of zero

Defining the operation for negative integer exponents

Defining the operation for fractional exponents

Laws of Exponents

Developing the additive law of exponents

Developing the law of repeated exponentiation

Graphing

Describing the graphs of exponential functions

Comparing graphs of exponential functions for different bases

Describing the graphs of logarithmic functions

Comparing graphs of logarithmic functions for different bases

Logarithms

Understanding the meaning of logarithms

Making connections between exponential and logarithmic equations

Scientific Notation

Converting numbers from ordinary notation to scientific notation, and vice versa

Developing principles for doing computations using scientific notation

Using the concept of order of magnitude in estimation

Progression

The unit begins with a brief introduction to the Alice metaphor. Next, students develop a set of rules for computing with exponents and generalize these rules to include zero, negative, and fractional exponents. Finally, students "undo" the exponential function to define the logarithmic function and learn about scientific notation. There are two POWs in this unit.

Who's Alice?

Extending Exponentiation

Curiouser and Curiouser!

Turning Exponents Around

Supplemental Activities

Unit Assessments

Pacing Guides

50-Minute Pacing Guide (20 days)

Day	Activity	In-Class Time Estimate
1	Who's Alice?	
	Alice in Wonderland	30
	POW 12: Logic from Lewis Carroll	15
	Homework: *Graphing Alice*	5
2	*Alice in Wonderland* (continued)	25
	Discussion: *Graphing Alice*	20
	Homework: *A Wonderland Lost*	5
3	Discussion: *A Wonderland Lost*	10
	Extending Exponentiation	
	Here Goes Nothing	30
	POW 12: Logic from Lewis Carroll (progress check)	10
	Homework: *A New Kind of Cake*	0
4	Discussion: *A New Kind of Cake*	10
	Piece After Piece	40
	Homework: *When Is Nothing Something?*	0
5	Discussion: *When Is Nothing Something?*	15
	Many Meals for Alice	35
	Homework: *In Search of the Law*	0
6	Discussion: *In Search of the Law*	15

	Having Your Cake and Drinking Too	30
	Homework: *Rallods in Rednow Land*	5
7	Discussion: *Rallods in Rednow Land*	15
	Having Your Cake and Drinking Too (continued)	30
	Homework: *Continuing the Pattern*	5
8	Discussion: *Continuing the Pattern*	45
	Homework: *Negative Reflections*	5
9	Presentations: *POW 12: Logic from Lewis Carroll*	15
	Discussion: *Negative Reflections*	5
	Curiouser and Curiouser!	
	A Half Ounce of Cake	30
	Homework: *It's in the Graph*	0
10	*A Half Ounce of Cake* (continued)	15
	Discussion: *It's in the Graph*	15
	POW 13: A Digital Proof	15
	Homework: *Stranger Pieces of Cake*	5
11	Discussion: *Stranger Pieces of Cake*	35
	POW 13: A Digital Proof (continued)	15
	Homework: *Confusion Reigns*	0
12	Discussion: *Confusion Reigns*	20
	All Roads Lead to Rome	25
	Homework: *Measuring Meals for Alice*	5
13	Discussion: *Measuring Meals for Alice*	15
	All Roads Lead to Rome (continued)	25

	POW 13: A Digital Proof (continued)	10
	Turning Exponents Around	
	Homework: Sending Alice to the Moon	0
14	Discussion: Sending Alice to the Moon	40
	Homework: Alice on a Log	10
15	Discussion: Alice on a Log	15
	Taking Logs to the Axes	30
	Homework: Base 10 Alice	5
16	Discussion: Base 10 Alice	30
	POW 13: A Digital Proof (work time)	10
	Homework: Warming Up to Scientific Notation	10
17	Discussion: Warming Up to Scientific Notation	10
	Big Numbers	40
	Homework: An Exponential Portfolio	0
18	Presentations: POW 13: A Digital Proof	20
	Discussion: An Exponential Portfolio	10
	Homework: "All About Alice" Portfolio	20
19	In-Class Assessment	40
	Homework: Take-Home Assessment	10
20	Exam Discussion	30
	Unit Reflection	20

90-minute Pacing Guide (12 days)

Day	Activity	In-Class Time Estimate
1	Who's Alice?	
	Alice in Wonderland	55
	POW 12: Logic from Lewis Carroll	25
	Homework: *Graphing Alice*	5
	Homework: *A Wonderland Lost*	5
2	Discussion: *Graphing Alice*	20
	Discussion: *A Wonderland Lost*	10
	Extending Exponentiation	
	Here Goes Nothing	30
	Piece After Piece	30
	Homework: *A New Kind of Cake*	0
	Homework: *When Is Nothing Something?*	0
3	Discussion: *A New Kind of Cake*	10
	Piece After Piece (continued)	10
	Discussion: *When Is Nothing Something?*	15
	Many Meals for Alice	35
	POW 12: Logic from Lewis Carroll (progress check)	10
	Homework: *In Search of the Law*	5
	Homework: *Rallods in Rednow Land*	5
4	Discussion: *In Search of the Law*	15

	Discussion: *Rallods in Rednow Land*	15
	Having Your Cake and Drinking Too	60
	Homework: *Continuing the Pattern*	0
5	Discussion: *Continuing the Pattern*	40
	Curiouser and Curiouser!	
	A Half Ounce of Cake	50
	Homework: *Negative Reflections*	0
6	Presentations: *POW 12: Logic from Lewis Carroll*	15
	Discussion: *Negative Reflections*	5
	Stranger Pieces of Cake	60
	POW 13: A Digital Proof	10
	Homework: *It's in the Graph*	0
7	*POW 13: A Digital Proof* (continued)	20
	Discussion: *It's in the Graph*	15
	All Roads Lead to Rome	50
	Homework: *Confusion Reigns*	5
8	Discussion: *Confusion Reigns*	20
	Measuring Meals for Alice	45
	POW 13: A Digital Proof (work time)	20
	Turning Exponents Around	
	Homework: *Sending Alice to the Moon*	5
9	Discussion: *Sending Alice to the Moon*	40
	Alice on a Log	45
	Homework: *Taking Logs to the Axes*	0

	Homework: *Base 10 Alice*	5
10	Discussion: *Taking Logs to the Axes*	10
	Discussion: *Base 10 Alice*	30
	Warming Up to Scientific Notation	40
	POW 13: A Digital Proof (work time)	10
	Homework: *An Exponential Portfolio*	0
11	Presentations: *POW 13: A Digital Proof*	20
	Big Numbers	40
	Discussion: *An Exponential Portfolio*	10
	Homework: *"All About Alice" Portfolio*	20
	Homework: *Take-Home Assessment*	0
12	*In-Class Assessment*	40
	Exam Discussion	30

Materials and Supplies

All IMP classrooms should have a set of standard supplies and equipment, and students are expected to have materials available for working at home on assignments and at school for classroom work. Lists of these standard supplies are included in the section "Materials and Supplies for the IMP Classroom" in *A Guide to IMP*. There is also a comprehensive list of materials for all units in Year 2.

Listed below are the supplies needed for this unit. General and activity-specific blackline masters are available for presentations on the overhead projector or for student worksheets. The masters are found in the *All About Alice* Unit Resources under *Blackline Masters*.

All About Alice

- Adding machine tape (3 or 4 rolls for the entire class)

More About Supplies

- Graph paper is a standard supply for IMP classrooms. Blackline masters of 1-Centimeter Graph Paper, ¼-Inch Graph Paper, and 1-Inch Graph Paper are provided so you can make copies and transparencies. (You'll find links to these masters in "Materials and Supplies for Year 2" [link] of the Year 2 guide and in the Unit Resources of each unit.)

Assessing Progress

All About Alice concludes with two formal unit assessments. In addition, there are many opportunities for more informal, ongoing assessments throughout the unit. For more information about assessment and grading, including general information about the end-of-unit assessments and how to use them, see "Assessment and Grading" in the *Year 2: A Guide to IMP* resource.

End-of-Unit Assessments

This unit concludes with in-class and take-home assessments. The in-class assessment is intentionally short so that time pressures will not affect student performance. Students may use graphing calculators and their notes from previous work when they take the assessments.

Ongoing Assessment

Assessment is a component in providing the best possible ongoing instructional program for students. Ongoing assessment includes the daily work of determining how well students understand key ideas and what level of achievement they have attained in acquiring key skills.

Students' written and oral work provide many opportunities for teachers to gather this information. Here are some recommendations of written assignments and oral presentations to monitor especially carefully that will offer insight into student progress.

- *Graphing Alice:* This assignment will give you information about how well students understand the basic Alice metaphor and about their comfort with nonlinear graphs.

- *Having Your Cake and Drinking Too:* This activity will reveal students' ability to work with the Alice metaphor in a complex situation.

- *Negative Reflections:* This assignment will tell you how well students understand the extension of exponentiation to negative exponents.

- *All Roads Lead to Rome:* This activity will give you information on students' ability to synthesize a variety of approaches to understanding a mathematical concept.

- *Alice on a Log:* This assignment will give you information on students' understanding of the basics about logarithms.

Supplemental Activities

All About Alice contains a variety of activities at the end of the student pages that you can use to supplement the regular unit material. These activities fall roughly into two categories.

- **Reinforcements** increase students' understanding of and comfort with concepts, techniques, and methods that are discussed in class and are central to the unit.

- **Extensions** allow students to explore ideas beyond those presented in the unit, including generalizations and abstractions of ideas.

The supplemental activities are presented in the teacher's guide and the student book in the approximate sequence in which you might use them. Below are specific recommendations about how each activity might work within the unit. You may wish to use some of these activities, especially the later ones, after the unit is completed.

***Inflation, Depreciation, and Alice* (reinforcement)** This activity provides other contexts for exponential growth and "shrinkage," or decay. To find formulas for Questions 1 and 2, students should recognize that increasing prices by 5% is the same as multiplying them by 1.05. They should reason similarly for depreciation. This activity can be assigned as soon as the basic idea of the Alice metaphor is clear.

***A Logical Collection* (reinforcement)** This activity contains several logic problems that make a good follow-up to the discussion of *POW 12: Logic from Lewis Carroll*.

***More About Rallods* (extension)** This activity builds on the situation from *Rallods in Rednow Land*, asking students to find a general formula for the sum of the first *n* powers of 2 and then having them explore other geometric sequences.

***Ten Missing Digits* (extension)** This activity expands the digit puzzle in *POW 13: A Digital Proof* to ten missing digits.

***Exponential Graphing* (reinforcement)** This activity offers students more opportunities to examine the graphs of exponential functions and can be assigned after *Stranger Pieces of Cake*.

***Basic Exponential Questions* (extension)** This activity raises some challenging questions about inequalities involving exponential expressions and can be used after *Stranger Pieces of Cake*. Question 2 is intentionally trivial (as it involves base 1). Question 3 follows up with a similar but more complicated problem. Question 4 is quite difficult to solve in general. The only whole-number solutions are the cases in which X is 2 and Y is 4, and vice versa. Students may find explanations for why there are no other solutions.

***Alice's Weights and Measures* (extension)** This activity explores issues of approximation. When a measurement is only an approximation, what effect does it have on computations that make use of that measurement? This exercise makes a good follow-up to *Measuring Meals for Alice,* in which students give only approximate answers.

***A Little Shakes a Lot* (reinforcement)** In this activity, students explore an interesting real-world use of logarithms: earthquakes. You can use this activity after *Sending Alice to the Moon*.

***Who's Buried in Grant's Tomb?* (extension)** This activity offers another setting in which students can explore the relationship between exponents and logarithms. This can be used after *Sending Alice to the Moon*.

***Very Big and Very Small* (extension)** This supplement asks students to identify and investigate more situations involving very big and very small numbers in contexts they find intriguing. It can be assigned following *Big Numbers.*

Who's Alice?

Intent

These introductory activities use excerpts from Lewis Carroll's book *Alice's Adventures in Wonderland* to introduce the context of and to lay the mathematical foundation for the unit.

Mathematics

Through a retelling of the classic story, eating ounces of cake and drinking ounces of beverage will serve as metaphors for exponential, or constant percentage, growth and decay, respectively. Students discover that eating C ounces of cake multiplies Alice's height by 2^C and drinking B ounces of beverage multiplies her height by $\left(\dfrac{1}{2}\right)^B$. They construct In-Out tables and graphs that display these data.

Finally, students examine a new context: a constant percentage decay in which the base is not $\dfrac{1}{2}$.

Progression

The first activity introduces the context for the unit and some important vocabulary. The other two activities introduce graphing and set the stage for more rigorous treatment of exponential functions. In addition, students begin work on the first POW of the unit, which focuses on logical reasoning.

Alice in Wonderland

POW 12: Logic from Lewis Carroll

Graphing Alice

A Wonderland Lost

Alice in Wonderland

Intent

This activity introduces the metaphor for exponential growth and decay that is used throughout the unit. Students begin thinking about how exponents work and establish basic exponential growth and decay expressions.

Mathematics

An excerpt from *Alice's Adventures in Wonderland* establishes a context for thinking about exponents, exponential expressions, and exponential functions. The metaphor will be used to motivate students and to help them extend the definition of exponentiation beyond whole-number exponents and understand laws relating to exponents. Students realize that eating C ounces of cake multiplies Alice's height by 2^C and drinking B ounces of beverage multiplies her height by $\left(\dfrac{1}{2}\right)^B$, eventually realizing this second multiplication process as equivalent to repeated division by 2.

Throughout the work with this metaphor, focus students' attention on the *factor* by which Alice grows rather than on the *amount* by which she grows. The cake and beverage create *multiplicative* changes, rather than *additive* changes, in Alice's height.

Progression

Working through some examples in groups, students consider *how much bigger* or *how much smaller* Alice grows and shrinks, and then they generalize the process. Class discussion then determines some agreements for interpretation of and notation for this metaphor. The activities *Graphing Alice* and *A Wonderland Lost* will strengthen the metaphor by linking the ideas of exponential growth to other representations and situations.

Approximate Time

55 minutes

Classroom Organization

Groups, followed by whole-class discussion

Materials

Cash register tape (optional)

Doing the Activity

Wait at least a day before discussing the mathematical topics to be studied in this unit, as doing so now may undermine the discovery that Alice's height is related to exponents.

Many students may be unfamiliar with Lewis Carroll's *Alice's Adventures in Wonderland,* so give a quick overview of the story to set the context, or ask a student to do so. To refresh your memory, the story begins with a young girl who dreams she spots a white rabbit who is carrying a pocket watch and muttering about how late he is. Alice follows him down a rabbit hole to a tea party with the Mad Hatter and goes on to have many other adventures. Although the cake and beverage are part of the original story, no specific numeric effect of eating or drinking is given.

Have the class read the excerpt and the paragraphs that follow it, but not the questions.

Acting Out "Doubling"
Before having students turn to the questions, it will be helpful to have them act out the concept of Alice's growth to give them a concrete sense of the effect of repeated doubling.

For example, you might use tiles on the floor to act out the doubling process. Have one student stand on the first tile, another on the second tile, another on the fourth tile, another on the eighth tile, and so on. To act out the halving process, you might give each group or pair of students a strip of cash-register tape, perhaps 20 feet long, and have them repeatedly fold it in half so they realize how quickly halving reduces the length of the strip.

After this introduction, have students work on the questions in their groups.

Nowhere in the problem are we told how tall Alice is to begin with; we know only by how much her height is multiplied. Some students may not like working with this kind of abstraction. You might suggest they make the situation more concrete by picking a particular height for Alice as a starting point, still focusing on the comparison between her starting height and her final height.

Discussing and Debriefing the Activity

Ask for volunteers to present each question as the rest of the class comments and elaborates. Given here is guidance on what needs to emerge from these presentations. As needed, remind students to focus on the *factor* by which Alice's height changes.

Eating C Ounces of Cake
Students should realize that if Alice eats 1 ounce of cake, then 1 more ounce, and so on, her height will double each time so that overall, if she eats C ounces, her height will be multiplied by 2^C. It is important that this fundamental generalization be very clear.

If necessary, have the presenter explain the specific cases in Question 1, and ask questions to bring out the use of an exponential expression in Question 2. **Is there another way to write 2 • 2? What if C 2s were multiplied together?**

If needed, remind students of the words **base** and **exponent** for referring to the number 2 and the C in the expression 2^C.

Note: When we say that eating C ounces of cake multiplies Alice's height by 2^C, we are assuming that eating C ounces all at once is the same as eating them one ounce at a time. This issue will be important in the discussion of the activity *Piece After Piece*. The generalization that eating C ounces of cake multiplies Alice's height by 2^C also incorporates the fact that if something is multiplied first by one factor (for example, *a*) and the resulting product is multiplied by another factor (for example, *b*), the original number has been multiplied altogether by *ab*. Thus, if the first ounce of cake doubles Alice's height and the second doubles her height again, altogether her height has been multiplied by 2 • 2, or 4. This is essentially the associative property of multiplication, as it states that $(h • 2) • 2 = h • (2 • 2)$, where *h* represents Alice's original height.

Drinking B Ounces of Beverage
Students typically find the beverage aspect of the situation somewhat more difficult. They need to realize that drinking B ounces of the beverage multiplies Alice's height by $\dfrac{1}{2^B}$, which they may initially write as $\left(\dfrac{1}{2}\right)^B$.

They are probably even more likely to recognize the change in Alice's height in terms of division rather than multiplication, dividing by 2^B. The unit will proceed more smoothly, however, if they focus on the question, **What has Alice's height been multiplied by?**

If students initially use division, help them make this transition, phrasing your questions in fairly explicit terms if necessary. **What is another way to express division by 2? How can you express this in terms of multiplication?** If students say 1 ounce of beverage multiplies Alice's height by 0.5, ask, **How else can you write 0.5?**

It is also important that students express this multiplication factor as a fraction rather than as a decimal, as this will make the pattern clear and help extend the operation of exponentiation, in upcoming activities, to include negative exponents.

The fact that drinking B ounces of beverage multiplies Alice's height by $\dfrac{1}{2^B}$ can be understood by students as the consequence of two basic ideas.

- Multiplying by $\dfrac{1}{2}$ repeatedly, B times, is the same as multiplying by $\left(\dfrac{1}{2}\right)^B$.

- Multiplying by $\left(\dfrac{1}{2}\right)^B$ is the same as multiplying by $\dfrac{1}{2^B}$. In other words,

$$\left(\frac{1}{2}\right)^B = \frac{1^B}{2^B} = \frac{1}{2^B}.$$

If students do not reason this way, you can introduce these ideas gradually, asking what Alice's height is multiplied by when she drinks a specific amount and then having students restate the result.

For example, if they say that drinking 3 ounces means Alice's height is multiplied by $\dfrac{1}{8}$, ask where the 8 comes from. It will be easier to pose this type of question if students write the multiplying factor as the fraction $\dfrac{1}{8}$ rather than the decimal 0.125.

Even after examples like this, many students will continue to interpret drinking multiple ounces of beverage as repeatedly dividing Alice's height by 2 rather than repeatedly multiplying it by $\dfrac{1}{2}$.

Key Questions

What has Alice's height been multiplied by?

Is there another way to write 2 • 2? What if C 2s were multiplied together?

What is another way to express division by 2? How can you express this in terms of multiplication?

How else can you write 0.5?

Supplemental Activity

Inflation, Depreciation, and Alice (reinforcement) provides other contexts for exponential growth and "shrinkage," or decay. To find formulas for Questions 1 and 2, students should recognize that increasing prices by 5% is the same as multiplying them by 1.05. They should reason similarly for depreciation.

POW 12: Logic from Lewis Carroll

Intent

This POW focuses on basic ideas of logic and deduction connected to the interests of Lewis Carroll, who was a mathematician as well as an author. This POW is more directed than usual, but the subject matter is likely to be new to many students, and the problem does have an open-ended component.

Mathematics

Logic and proof are at the core of the mathematician's work. This POW explores situations of logical deduction and, in particular, syllogisms. Technically, a *syllogism* is an argument in which a conclusion is drawn based on two premises. For example, the premises "Socrates is a man" and "All men are mortal" lead to the conclusion "Socrates is mortal." Students often spend considerable time studying formal logic in a traditional geometry course. This POW asks them to use this type of reasoning in a less formal setting.

Progression

Students are introduced to this POW early in the unit, with time spent to help them clarify the problem as well as to confer with one another about their deductions. Presentations follow about a week later.

Approximate Time

15 minutes for introduction
1 to 2 hours for activity (at home)
10 minutes for progress check
15 minutes for presentations

Classroom Organization

Groups, then individuals, followed by whole-class presentations

Doing the Activity

Introduce the activity and discuss the two examples. Point out that the text is not claiming that any of the individual statements is actually true. Instead, the focus is on what conclusions could be drawn *if* they were true.

If possible, provide some in-class time a few days after students have begun work on this POW as a progress check of sorts. You might give them five or ten minutes in their groups to compare ideas and then take general questions. Or have students discuss the first question or two either in groups or as a class.

You might also offer these additional examples from Lewis Carroll to help students understand the POW, or let groups spend some time trying to make up more problems of this type.

a. No quadrupeds can whistle.
b. Some cats are quadrupeds.
Conclusion: "Some cats can't whistle."

a. All clever people are popular.
b. All obliging people are popular.
No new conclusion can be drawn.

a. All puddings are nice.
b. This dish is a pudding.
c. No nice things are wholesome.
Conclusion: "This dish is not wholesome."

The day before the POW is due, identify three students to prepare presentations.

Discussing and Debriefing the Activity

One approach for presentations and discussion is to have the three students alternate presentations on the six questions. After each question, let the other presenters add any comments and then ask the rest of the class for comments. Here are some things to look for.

1. Conclusion: "Senna is not nice." As this is a straightforward example, make sure everyone understands why this conclusion is valid.

2. No new conclusion can be drawn. You might mention that shillings are coins formerly used in Britain (though this is not relevant to the problem). Make sure all students understand that they cannot conclude, "These coins are shillings."

3. Conclusions: "Some wild pigs are fat" or "Some fat pigs are wild." Students may not recognize these as legitimate conclusions. If they decided there is no possible new conclusion, that's okay; it's a matter of opinion whether these conclusions are "new." But they should realize that both of these conclusions must be true, as is the conclusion "There are some fat, wild pigs."

4. No new conclusion can be drawn. Students may suggest as a conclusion "Some unprejudiced persons are liked." According to formal logic, this doesn't follow from statement b, although everyday language usage suggests that if we say "some are" we also are saying "some are not." This is worth discussing if it arises. Point out that we often draw conclusions from things people say that are not necessarily inherent in what they say.

5. Main conclusion: "Babies cannot manage a crocodile." Other, subsidiary conclusions use only two of the three statements, such as "Babies are despised" (from a and c) and "Illogical persons cannot manage a crocodile" (from b and c). Discuss how the main conclusion can be drawn by combining one of these subsidiary statements with the remaining original statement.

6. Main conclusion: "No bird in this aviary lives on mince pies." This conclusion uses all four statements. Subsidiary conclusions use only two of the statements: "All my birds are ostriches" (from a and d), "No birds in this aviary are less than 9 feet tall" (from b and d), and "No birds that are 9 feet tall live on mince pies" (from a and c). Try not to get caught up in such technicalities as whether ostriches are *exactly* 9 feet tall or *at least* 9 feet tall. Also, help students to realize that none of the statements guarantees that there are any birds in the aviary or that "I" have any birds. In general, a statement like "All *x*'s are *y*" doesn't imply that there are any *x*'s.

Ask presenters, and perhaps volunteers, to give one of the examples they made up for Part II of the POW for the class to work on.

Supplemental Activity

A Logical Collection (reinforcement) contains several logic problems that make a good follow-up to the POW discussion.

Graphing Alice

Intent

This activity will strengthen students' understanding of exponential relationships and, as they consider larger domains, their curiosity about what more there is to know.

Mathematics

Students examine graphs of base 2 and base 3 exponential relationships and confront the dilemmas posed by scaling exponential functions. They also examine the similarities and differences among graphs of varying bases, including different natural numbers and unit fractions. The nonlinearity of the relationships is emphasized. The activity also sets the stage for examining exponential functions with domains extended beyond the natural numbers.

Progression

This activity makes a good homework assignment, with class discussion occurring after work on the activity *Alice in Wonderland* is completed. The discussion concludes by raising questions about negative and zero inputs and emphasizing nonlinear growth by introducing the concepts **absolute growth** and **percentage growth.**

Approximate Time

5 minutes for introduction
20 minutes for activity (at home or in class)
20 minutes for discussion

Classroom Organization

Individuals, then groups, followed by whole-class discussion

Doing the Activity

Read through the introduction as a class. Discuss why students might want the scales on the two axes to be different.

Discussing and Debriefing the Activity

Have students share their findings in their groups. You might have them make chart-paper versions of their work in Questions 1 to 4 so all four graphs can be easily compared during the class discussion of Question 5.

Note: Students are expected only to plot individual points for the graphs in Questions 1 to 4, using specific positive integer values for *x,* rather than to graph the general functions. As the unit progresses, they will gradually move toward creating complete graphs of such functions as $y = 2^x$.

Question 1

The main focus for the discussion of Question 1 will likely be the scaling of the axes. One error many students make is scaling the y-axis so that the values 2^1, 2^2, 2^3, 2^4, and so on are equally spaced, thus creating a linear graph. If the presenter for Question 1 makes this mistake, ask a general question about scaling to bring this out, such as, **Is the difference between 2^2 and 2^3 the same as the difference between 2^3 and 2^4?**

When the graph is scaled correctly, students should realize that Alice's height does not grow linearly, as the points do not lie on a straight line.

Ask students to express the rule for the graph using function notation. **Using f to represent the function, how can you write an equation for this graph using function notation?** Using x for the independent variable, they will probably write $f(x) = 2^x$.

After the graph is finished, post it on chart paper, labeling it appropriately. Students can add to the graph as the definition of exponentiation is extended during the unit.

You may need to emphasize that the scales for the x- and y-axes can be different. In fact, for this question, if the scales are the same, the graph will be very tall in comparison to its width.

Question 2

The issues that arise for Question 2 will likely be similar to those for Question 1. Students will need to use a different scale for the y-axis than was used for Question 1.

Ask students to express this relationship using function notation and a new letter for the function name. For instance, they might write $g(x) = \left(\dfrac{1}{2}\right)^x$.

Questions 3 and 4

For these questions, elicit generalizations analogous to those developed in the discussion of *Alice in Wonderland*. That is, if Alice eats C ounces of the new cake, her height is multiplied by 3^C. If she drinks B ounces of the new beverage, her height is multiplied by $\dfrac{1}{3^B}$.

Introduce the terms *base 3 cake* and *base 3 beverage* for the new cake and beverage, and have a volunteer explain why these terms apply. Ask students what they would call the cake and beverage in the original problem. They should recognize that the appropriate terms are *base 2 cake* and *base 2 beverage.* Explain that, unless otherwise indicated, they should assume the cake and beverage in "Alice problems" are base 2 cake and base 2 beverage and that, whatever type is used, the beverage and the cake within a given problem will always use the same base.

The graphs for Questions 3 and 4 will require even more difference in the scales than those for Questions 1 and 2. Have students express these functions using function notation.

Question 5
You might let volunteers share their observations about Question 5 and have the rest of the class comment. There are several comparisons students can make.

The main observation is that in the cake problems (Questions 1 and 3), the *y*-value goes up rapidly as the *x*-value increases, while in the beverage problems (Questions 2 and 4), the *y*-value seems to get closer to 0 as the *x*-value increases. Students may also point out that the larger the base, the more extreme the change in *y* as *x* changes.

Extending the Graphs to the Left
As a lead-in to later discussion of the use of zero or negative integers as exponents, ask groups to discuss where each graph would cross the *y*-axis and what each graph might look like for negative inputs. Where would each graph cross the *y*-axis? What will happen to the graphs as *x* becomes negative? These issues should be considered only briefly at this point, in terms of what extending the graphs might suggest rather than in terms of what zero or negative exponents might mean.

After a few minutes of small-group discussion, let volunteers share ideas. Students may notice that all the graphs seem likely to cross the *y*-axis at the point (0, 1). They may also realize that in the graphs from Questions 1 and 3, the *y*-values seem likely to get closer to 0 as *x* becomes more negative, while in the graphs from Questions 2 and 4, the *y*-values seem likely to get larger and larger as *x* becomes more negative.

Absolute Growth Versus Percentage Growth
Now ask students to focus on the case of base 2 cake, and suggest they have Alice start from a specific height, such as 5 feet. Ask, How does the effect on Alice's height of her third ounce of cake compare to the effect of her fifth ounce of cake?

Using a starting height of 5 feet, they might realize that the third ounce of cake increases Alice's height from 20 feet to 40 feet while the fifth ounce increases her height from 80 feet to 160 feet. Bring out that although the latter is a greater increase (80 versus 20 feet of growth), both cases involve doubling her height. Introduce the terms **percentage growth** for the proportional rate of increase (found by dividing the final value by the initial value) and **absolute growth** for the numeric difference (found by subtracting the initial value from the final value).

Key Questions

Is the difference between 2^2 and 2^3 the same as the difference between 2^3 and 2^4?

Using f to represent the function, how can you write an equation for this graph using function notation?

Where would each graph cross the y-axis? What will happen to the graphs as x becomes negative?

How does the effect on Alice's height of her third ounce of cake compare to the effect of her fifth ounce of cake?

A Wonderland Lost

Intent

This activity offers students a real-world context for the phenomenon of exponential decay.

Mathematics

Students examine an exponential decay function expressed as a constant percentage decrease, beginning from a context and deriving the tabular, graphical, and symbolic representations of the associated exponential function. Translating a percentage decrease into a symbolic rule, possibly in standard exponential form, will be challenging.

Progression

Students work on this task individually and share results with the whole class. The class will return to this graph in the next activity, *Here Goes Nothing.*

Approximate Time

5 minutes for introduction
20 minutes for activity (at home or in class)
10 minutes for discussion

Classroom Organization

Individuals, then groups, followed by whole-class discussion

Doing the Activity

Although this is an individual task, students will benefit from some initial whole-class or small-group work. After they read the task, ask them to describe the situation in their own words.

Help them to clarify that the 10 percent in the problem is always a percentage of the remaining forest, which decreases each year. You might ask, *If 10 percent of the forest is destroyed each year, how many years will it take until the forest is completely gone? Ten years, right?*

Some students will recognize that this statement is incorrect. Getting someone who understands the concept to explain why should clear up any possible confusion. If needed, introduce a simpler specific area for the rain forest, such as 100,000 square miles, for students to work with.

Discussing and Debriefing the Activity

Give students time to share ideas and ask questions in their groups before beginning the discussion.

Questions 1 and 2

Have students give their numeric results, year by year, for Questions 1 and 2. You might record their results in an In-Out table, including the initial area given in the activity as the second coordinate of a point whose first coordinate is 0.

Have someone suggest scales for a graph of this information. You might sketch a graph of the table data or have another student share his or her graph.

Questions 3 and 4

Students are likely to have had difficulty developing a general rule for this situation; repeatedly subtracting 10 percent does not lend itself to a simple expression. The key to obtaining a general rule is to recognize that decreasing something by 10 percent is the same as multiplying it by 0.9.

You might ask, **How does the area at the end of each year compare to the area at the beginning of that year?** The relationship may be clearer if you have students consider a round number, such as 100,000 square miles, and ask how the area after a year compares to the initial area.

Another approach is to have them write the initial area as A and the new area as $A - 0.1A$, and then combine terms to get $0.9A$.

With help, students should be able to formulate the general expression $1{,}200{,}000 \bullet 0.9^x$ for the area after X years.

Finally, discuss how the rain forest problem relates to Alice. Students should figure out that the situation is essentially the same as that of Alice's beverage, except that Alice's height decreases by 50% per ounce while the rain forest decreases by 10% per year.

Key Questions

If 10 percent of the forest is destroyed each year, how many years will it take until it is all gone? Ten years, right?

How does the area at the end of each year compare to the area at the beginning of that year?

What is the current area multiplied by to figure the area for the next year?

Extending Exponentiation

Intent

In these activities, students will derive several rules for computing with exponents and extend their understanding of exponential expressions to include zero and negative integers.

Mathematics

In these activities, students will use the Alice metaphor and patterns in lists like $2^3 = 8$, $2^2 = 4$, and $2^1 = 2$ to derive a number of rules for working with exponents.

$$A^x \bullet A^y = A^{x+y}$$

$$\left(A^x\right)^y \bullet \left(A^y\right)^x = A^{xy}$$

$$A^0 = 1 \text{ if } A \neq 0$$

$$A^{-x} = \frac{1}{A^x} \text{ if } A \neq 0$$

Progression

The activities first give meaning to an exponent of zero and then develop the **additive law of exponents** and the **law of repeated exponentiation.** They conclude with the extension of exponents to include negative integers. In addition, students will present their work on the first POW of the unit.

Here Goes Nothing

A New Kind of Cake

Piece After Piece

When Is Nothing Something?

Many Meals for Alice

In Search of the Law

Having Your Cake and Drinking Too

Rallods in Rednow Land

Continuing the Pattern

Negative Reflections

Here Goes Nothing

Intent

Students will consider zero as the value of an exponent and use patterns, tables, and graphs to explain their findings.

Mathematics

Students consider the meaning of 2^0, initially through the Alice metaphor, and focus on the idea that multiplying by the quantity 2^0 must be equivalent to multiplying by 1, the multiplicative identity. In addition to rationalizing the definition that $2^0 = 1$ through the Alice situation, they look at the patterns of powers of 2 in graphical and tabular arguments.

Progression

Students collaborate in groups to consider the impact on Alice's height of eating no cake and what that implies for the meaning of 2^0. The follow-up discussion emphasizes $2^0 = 1$ as a mathematical definition. Later activities will extend this definition to other bases.

Approximate Time

30 minutes

Classroom Organization

Groups, followed by whole-class discussion

Doing the Activity

Before students begin, you may want to review the formula developed earlier, in which eating C ounces of cake multiplies Alice's height by a factor of 2^c.

For Question 3, you may need to clarify that groups only need to substitute 0 for C in the expression 2^C—they don't need to evaluate the expression.

Discussing and Debriefing the Activity

Have volunteers from different groups answer Questions 1 to 4. They should be clear that eating 0 ounces of cake doesn't change Alice's height, which means her height is multiplied by 1.

The presenter for Question 2 might refer to the posted graph (from Question 1 of *Graphing Alice*) and show that if the graph is continued to the left, it might hit the y-axis at $y = 1$. Although the graph might not point clearly to a y-value of 1, at this time students need only to realize that this result is reasonable.

For Question 3, they should recognize that substituting 0 for C in the expression 2^C gives 2^0.

For Question 4, you may need to summarize what was said in Questions 1 to 3 so that the appropriate conclusion becomes clear: that it seems to make sense for 2^0 to be equal to 1.

Back to "A Wonderland Lost"

Reinforce the discussion by returning to the graph from students' work on *A Wonderland Lost.* Remind students of the rule they found for the rain forest situation $(1{,}200{,}000 \cdot 0.9^x)$ and ask what value they want when $X = 0$. The graph should include the point (0, 1,200,000), which shows they want $1{,}200{,}000 \cdot 0.9^x$ to equal 1,200,000 when $X = 0$, meaning they want 0.9^0 to be equal to 1.

$2^0 = 1$ Is a Definition

It's very important that students realize that a definition is needed to give meaning to the expression 2^0. Bring out that it makes sense to say "2^3 means to multiply three 2s together" and "2^5 means to multiply five 2s together," but it doesn't make sense to say "2^0 means to multiply zero 2s together." So a decision has to be made—there needs to be a convention, or agreement, as to the value of 2^0.

Tell students that long ago, mathematicians agreed to *define* 2^0 as having a value of 1. The purpose of this activity is to show that this is the *most reasonable* definition, because it fits what happens to Alice and it fits the graph. (You can expect students both to resist and to forget this definition. The notion that any computation involving multiplication and zero gives a result of zero is a strong one, and it may take some students a while to let go of this idea.)

The Exponential Pattern

Use the pattern of powers of 2 to reinforce the idea that $2^0 = 1$. Make a list like this.

$$2^5 = 32$$
$$2^4 = 16$$
$$2^3 = 8$$
$$2^2 = 4$$
$$2^1 = 2$$

Get the class to articulate this pattern in various ways, such as "Each result is twice the one below it" or "Divide by 2 as the exponent goes down by 1."

What would be the next equation in this pattern? Students should recognize that the natural way to continue this pattern is with the equation $2^0 = 1$.

Remind students that $2^0 = 1$ is, ultimately, a definition. The usual definition of exponents in terms of repeated multiplication has broken down, because there are no 2s to multiply. So we must use some other method of defining the expression, and it makes sense to formulate the definition in a way that is consistent with other ideas.

What are the various reasons the definition "$2^0 = 1$" makes sense? Students should identify at least three reasons.

- It fits the rule that 2^C tells what to multiply Alice's height by when she eats C ounces of cake.
- It seems a reasonable way to extend what they already have of the graph of the equation $y = 2^x$.
- It fits the pattern of powers of 2.

As students realize that all these methods agree, they should become more satisfied with the definition. Soon they will begin developing the **additive law of exponents** and will find that this law provides another way to justify the definition.

Key Questions

What would be the next equation in this pattern?

What are the various reasons the definition "$2^0 = 1$" makes sense?

A New Kind of Cake

Intent

Students check their understanding of using zero as an exponent by extending their work from base 2 to other bases. In addition, they look at graphs of $y = 2^x$ and $y = x^2$ to explore the effects of switching base and exponent.

Mathematics

The activity provides review and reinforcement of ideas about the equation and graph of the exponential function and about zero as an exponent. In this instance, students use base 5. They also compare the graphs of $y = 2^x$ and $y = x^2$ with the primary intention of recognizing that the expression 2^x grows much faster than x^2 as x increases.

Progression

After some work on their own, students come together as a class to compare results. This activity transitions to the next activity, *Piece After Piece,* by focusing on the question "What is Alice's height multiplied by?" after she eats various amounts of cake.

Approximate Time

20 minutes for activity (at home or in class)
10 minutes for discussion

Classroom Organization

Individuals, followed by whole-class discussion

Doing the Activity

Introduce this activity by telling students they will now consider what happens with another base.

Discussing and Debriefing the Activity

You may want to have a volunteer present his or her ideas on Part I. The presenter should note that eating 0 ounces multiplies Alice's height by 1 and that this does seem to fit the graph obtained in Question 2. Because the scale on that graph covers such a wide range of values, however, this is not a very convincing argument for defining 5^0 as 1.

The key element of Question 3 is part c. The student might argue, for instance, that eating 2 ounces multiplies Alice's height by 25, which is 5^2, so eating 0 ounces should multiply her height by 5^0, which means 5^0 must equal 1.

The presenter for Question 4 should be able to use reasoning analogous to that in the discussion of powers of 2 from *Here Goes Nothing* to argue that it makes sense to define 5^0 as 1.

For Part II, students will presumably have graphed the two equations by plotting individual points. The main idea that needs to emerge is that the expression 2^x grows much faster than the expression x^2 as x increases. If students did not realize this, they probably did not take their graphs far enough out to the right. You might have the class choose some numbers between 10 and 20 and find the values of the expressions 2^x and x^2 for comparison, or graph the equations on a calculator.

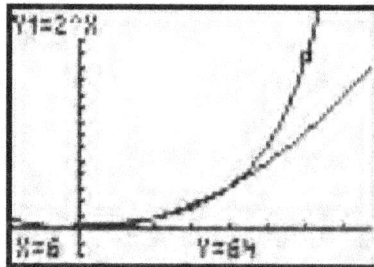

Piece After Piece

Intent

Students use the metaphor of eating cake and drinking beverage to explore the additive law of exponents.

Mathematics

In the unit's opening activity, *Alice in Wonderland,* students developed the principle that eating C ounces of cake multiplies Alice's height by 2^C. This generalization assumes that eating C ounces is the same as eating 1 ounce C times—in other words, it doesn't matter whether Alice eats her cake all at once or one ounce at a time. Now students will make explicit use of this assumption to develop the additive law of exponents. The **additive law of exponents** says that when two exponential expressions with the same base are multiplied, this property holds:

$$A^X \bullet A^Y = A^{X + Y}$$

Progression

Working in small groups, students use the Alice metaphor to develop the additive law of exponents for base 2 cake and beverage. In the follow-up discussion, they use repeated multiplication to explain the additive law of exponents for base 2, and then they generalize the procedure. The next activity, *When Is Nothing Something?,* brings together their ideas about zero as an exponent and the additive law of exponents.

Approximate Time

40 minutes

Classroom Organization

Groups, followed by whole-class discussion

Doing the Activity

The real question of this task—What happens if Alice doesn't eat all her cake in one sitting?—is suggested in the first sentence.

Have students read the activity. Encourage some conjecture, and then have them explore the task in their groups.

Discussing and Debriefing the Activity

Begin by having students answer Question 1. The presenter for Question 1a might describe what happens overall, saying something like, "Her height is multiplied by 256." If so, ask for an explanation of how the presenter arrived at this conclusion. What are the stages of her height change? Ensure that everyone understands that Alice's height is multiplied first by 8 and then by 32. Bring out that 8 comes from the expression 2^3 and that 32 comes from the expression 2^5.

You might suggest students represent Alice's original height with a variable. **If Alice's initial height is *h* and she eats 3 ounces of cake, what will her new height be?** Write the response as $h \cdot 2^3$. Proceed similarly with the second stage, writing the next result as $(h \cdot 2^3) \cdot 2^5$. Then ask a similar question for the 8-ounce piece of cake.

If students used a specific initial height, they will certainly recognize that the overall result is the same as eating 8 ounces of cake. Focus on the explanation for this in terms of the arithmetic. **How does the arithmetic explain why the results are the same?** A reasonable justification might be, "Multiplying by 8 and then by 32 is the same as multiplying by 256."

The goal is to bring out what this means in terms of expressions with exponents. Essentially, it says that $2^3 \cdot 2^5$ is the same as 2^8.

This fact may seem obvious to some students but not to others. A good way to clarify the relationship is to make the individual factors of 2 explicit by asking, for instance, where the factor of 8 comes from. Return to the expression 2^3, and ask students to break it down into individual factors—that is, as $2 \cdot 2 \cdot 2$. Proceeding similarly with the factor 2^5, they should determine that $2^3 \cdot 2^5$ can be written as

$$(2 \cdot 2 \cdot 2) \cdot (2 \cdot 2 \cdot 2 \cdot 2 \cdot 2)$$

At the same time, they should note that the single expression 2^8 is equal to

$$2 \cdot 2 \cdot 2 \cdot 2 \cdot 2 \cdot 2 \cdot 2 \cdot 2$$

You can ask about the number of 2s in each expression—**How many 2s are in the expression $2 \cdot 2 \cdot 2$? In $2 \cdot 2 \cdot 2 \cdot 2 \cdot 2$? In $2 \cdot 2 \cdot 2 \cdot 2 \cdot 2 \cdot 2 \cdot 2 \cdot 2$?**—and then return to the exponential forms of these expressions to develop the equation

$$2^3 \cdot 2^5 = 2^8$$

Question 2

This is a good time to look at some other examples. Ask for volunteers to share their work on Question 2, and conduct a brief version of the development above for one or two examples.

If students offer examples with small exponents, have the class write out the individual factors, as done here.

$$2^2 \bullet 2^3 = 2^5$$

$$(2 \bullet 2) \bullet (2 \bullet 2 \bullet 2) = 2 \bullet 2 \bullet 2 \bullet 2 \bullet 2$$

Move from examples like this one, in which students can actually count the factors, to examples in which they simply add the exponents, such as

$$2^{17} \bullet 2^{32} = 2^{49}$$

Finally, ask for a generalization of this process. Students will probably be able to develop an equation such as

$$2^X \bullet 2^Y = 2^{X+Y}$$

Question 3

Have a volunteer present Question 3. If it doesn't arise in the presentation, ask explicitly how the reasoning would apply if the situation were about beverage instead of cake. Students should be able to develop a "beverage counterpart" to the last equation along the lines of

$$\left(\frac{1}{2}\right)^X \bullet \left(\frac{1}{2}\right)^Y = \left(\frac{1}{2}\right)^{X+Y}$$

The Additive Law of Exponents

As an additional stage in the development of the general **additive law of exponents,** ask students to make up similar examples of what would happen if Alice consumed another type of cake. For example, ask, Does eating 3 ounces and then 5 ounces of base 7 cake have the same effect as eating 8 ounces of that cake? Why? Students should be able to explain this with an expression such as

$$\underbrace{\left(7 \bullet 7 \bullet 7\right)}_{\text{3 factors}} \bullet \underbrace{\left(7 \bullet 7 \bullet 7 \bullet 7 \bullet 7\right)}_{\text{5 factors}}$$

pointing out that this gives a total of 8 factors, so $7^3 \cdot 7^5$ is equal to 7^8.

Depending on how students respond, you might go directly from such examples to the most general case, or you might have them develop generalizations for specific bases other than 2. For example, they might come up with the equation

$$7^X \bullet 7^Y = 7^{X+Y}$$

and generalize to derive the principle

$$A^X \bullet A^Y = A^{X+Y}$$

Help students understand this visually by encouraging them to write each exponential expression as a product of a group of factors. **What does A^X mean? What does A^Y mean? How would you write these expressions without using exponents?** Match the individual exponential expressions with their "written out" forms to get a display like this.

Identify the generalization $A^X \bullet A^Y = A^{X+Y}$ as the **additive law of exponents.** Point out that it involves a situation in which two things hold true:

- The bases are the same.
- The two exponential expressions are being multiplied.

Post the additive law of exponents together with the explanation.

Key Questions

What are the stages of Alice's height change?

If Alice's initial height is h and she eats 3 ounces of cake, what will her new height be?

How does the arithmetic explain why the results are the same?

How many 2s are in the expression $2 \bullet 2 \bullet 2$? In $2 \bullet 2 \bullet 2 \bullet 2 \bullet 2$? In $2 \bullet 2 \bullet 2 \bullet 2 \bullet 2 \bullet 2 \bullet 2 \bullet 2$?

Does eating 3 ounces and then 5 ounces of base 7 cake have the same effect as eating 8 ounces of that cake? Why?

What does A^X mean? What does A^Y mean? How would you write these expressions without using exponents?

When Is Nothing Something?

Intent

Students focus on the special meaning of zero as an exponent and then return to the additive law of exponents to explain zero as an exponent in one more way.

Mathematics

Zero is commonly referred to as "nothing" but rarely in mathematics does zero have no meaning. Zero has two general uses in mathematics: as a placeholder in our place-value number system and as the real number 0, located halfway between −1 and 1 on the number line.

Progression

Students work on this task individually. The brief follow-up discussion connects students' methods of justifying their understanding of zero as an exponent to the **additive law of exponents.**

Approximate Time

20 minutes for activity (at home or in class)
15 minutes for discussion

Classroom Organization

Individuals, followed by whole-class discussion

Doing the Activity

As a reminder, ask the class what 2^0 has been defined to equal. Tell students that in this activity, they will explore zero as an exponent in more depth, as well as consider when zero means something and when it means "nothing."

Discussing and Debriefing the Activity

Have several volunteers offer their explanations for Question 1. Questions 2 and 3 give students an opportunity to be both imaginative and reflective. There is no specific mathematical content that needs to emerge here, so students can shape the discussion by what they have to offer.

Zero as an Exponent in the Additive Law of Exponents

If the topic hasn't yet surfaced, ask, How could you use 2^0 in the additive law of exponents? Ask for an instance of the additive law of exponents that uses 2^0. Suppose, for example, students suggest the equation

$$2^0 \bullet 2^3 = 2^3$$

Ask, What number does 2^0 act like in this equation? To clarify this question, replace each 2^3 in the equation by the number 8 and the 2^0 by a box, as shown here, and ask, **What number should we put in the box to give a true equation?**

$$2^0 \bullet 2^3 = 2^3$$

$$\downarrow \qquad \downarrow \qquad \downarrow$$

$$\square \bullet 8 = 8$$

Students should realize that the missing number must be 1, so 2^0 is acting like 1 in the equation $2^0 \bullet 2^3 = 2^3$.

Bring out that this analysis, based on the additive law of exponents, is yet another reason to define 2^0 as 1. Emphasize that this is consistent with the other three explanations students have found to justify this definition:
- Alice eating no cake
- The graph of $y = 2x$
- The pattern of exponential values

Key Questions

How could you use 2^0 in the additive law of exponents?

What number does 2^0 act like in this equation?

What number should we put in the box to give a true equation?

Many Meals for Alice

Intent

Students use the metaphor of eating cakes and drinking beverages to explore another general law about exponents, the law of repeated exponentiation.

Mathematics

When an exponential expression is raised to a power, the following rules hold:.

$$\left(A^D\right)^M = \left(A^M\right)^D = A^{DM}$$

Students use the Alice metaphor to derive and justify these rules, known as the **law of repeated exponentiation.**

Progression

Students work in small groups and then as a class to examine what happens when Alice eats several meals of the same size.

Approximate Time

35 minutes

Classroom Organization

Groups, followed by whole-class discussion

Doing the Activity

Refresh students' memories about the activity *Piece After Piece,* in which they used the Alice metaphor to help derive a general law for exponents, and mention that they will discover another law of exponents today.

If groups have trouble generalizing from Question 2, you might suggest other specific examples for them to explore.

The main idea of this activity can be brought out based on students' work in Question 2; Questions 3 through 5 are primarily for groups who work more quickly. You may want to wait until all groups have at least begun Question 3 before starting the discussion.

Discussing and Debriefing the Activity

Have one or two students give their results for Question 2, referring to Question 1 as needed.

Students may express their answers to Question 1 as powers of 8, because each meal multiplies Alice's height by 2^3, or 8. In general, this gives 8^M as the multiplying factor for M meals of 3 ounces each.

For the general expression in Question 2, they will likely replace 8 by 2^D and get $\left(2^D\right)^M$ as the factor by which her height is multiplied after M meals of D ounces each.

Help the class obtain a general equation for simplifying exponential expressions by comparing two approaches for Question 2. In the approach just described, Alice ate M meals of D ounces each. A second approach is to recognize that Alice is eating of total of DM ounces, so her height is multiplied by 2^{DM}. If no one uses this second approach, ask such questions as, **How much cake was eaten altogether? What does that do to Alice's height?** For Question 1, the answers are "3M ounces" and "multiply by 2^{3M}." For Question 2, the answers are "DM ounces" and "multiply by 2^{DM}."

After establishing that the sequence in which the cake is eaten doesn't matter, comparing the two methods gives

$$\left(2^D\right)^M = 2^{DM}$$

Will this rule work with bases other than 2? Students will probably be able to generalize the rule as

$$\left(A^D\right)^M = A^{DM}$$

Once students have developed this general formula, ask them to explain the process in terms of repeated multiplication. **Can you write the process as a long-multiplication problem?** They might use a sequence of equalities such as those listed here to connect the two sides of the general equation.

$$\left(A^D\right)^M = \underbrace{\left(A^D \bullet A^D \bullet \ldots \bullet A^D\right)}_{M \text{ factors}}$$

$$= \underbrace{\underbrace{\left(A \bullet A \bullet \ldots \bullet A\right)}_{D \text{ factors}} \bullet \underbrace{\left(A \bullet A \bullet \ldots \bullet A\right)}_{D \text{ factors}} \bullet \underbrace{\left(A \bullet A \bullet \ldots \bullet A\right)}_{D \text{ factors}}}_{M \text{ factors}}$$

$$= \underbrace{\left(A \bullet A \bullet A \bullet \ldots \bullet A\right)}_{DM \text{ factors}}$$

$$= A^{DM}$$

Post the general formula

$$\left(A^D\right)^M = A^{DM}$$

with an explanation like the preceding one , and label it the **law of repeated exponentiation.**

If time allows, use Questions 3 to 5 to develop the rule

$$\left(2^D\right)^M = \left(2^M\right)^D$$

Students might explain this principle using the fact that both expressions are equal to 2^{DM} by the reasoning in Question 2.

Key Questions

How much cake was eaten altogether? What does that do to Alice's height?

Will this rule work with bases other than 2?

Can you write the process as a long-multiplication problem?

In Search of the Law

Intent

In *Piece After Piece* and *Many Meals for Alice,* students derived general rules for operations involving exponential expressions using the Alice metaphor. They now use these rules to derive new ones.

Mathematics

In this activity, students derive the rules listed here by applying the Alice metaphor, the additive law of exponents, and the law of repeated exponentiation.

$$A^X \bullet B^X = (A \bullet B)^X$$

$$A^X \bullet A^X = A^{2X}$$

Progression

Students work on the activity individually. The follow-up discussion brings out justifications for several general principles of exponents and establishes the convention of calling 0^0 *undefined.* In the activity *Confusion Reigns,* students will reexamine these general principles so that they don't simply memorize rules.

Approximate Time

20 minutes for activity (at home or in class)
15 minutes for discussion

Classroom Organization

Individuals, then groups, followed by whole-class discussion

Doing the Activity

Students have developed several rules for exponents. Tell them that they could derive many, many such rules. The rules identified so far use only one base. In this activity, students will look for rules that work when the base changes.

Discussing and Debriefing the Activity

Let students briefly collaborate and compare ideas in their groups, and then have them report. Focus the discussion on the explanations of what is happening with the factors in a given exponential expression. In the long run, having students understand specific examples will be more productive than having them memorize formulas.

Question 1

For Question 1, students should know how to regroup and pair the factors. If the presenter doesn't bring this out clearly, encourage the audience to ask for clarification. For instance, if the presenter is working with the example $3^7 \bullet 5^7$ given in the problem, drawing out a variety of ways to record this as repeated multiplication can yield patterns that suggest a "law." **In what ways can we write $3^7 \cdot 5^7$ as repeated multiplication?** Answering this should yield expressions such as these.

$$3 \bullet 3 \bullet 3 \bullet 3 \bullet 3 \bullet 3 \bullet 3 \bullet 5 \bullet 5 \bullet 5 \bullet 5 \bullet 5 \bullet 5 \bullet 5$$

$$(3 \bullet 5) \bullet (3 \bullet 5) \bullet (3 \bullet 5) \bullet (3 \bullet 5) \bullet (3 \bullet 5) \bullet (3 \bullet 5) \bullet (3 \bullet 5)$$

If students don't suggest this second example, encourage them to realize that they can pair up and rearrange the factors by asking, **How might the fact that there are the same number of 3s as 5s be useful?** Students should notice that the expression $3^7 \bullet 5^7$ can be rewritten as $(3 \bullet 5)^7$.

If needed, have another student present a similar example. Work toward a generalization of the principle that any product of the form $A^X \bullet B^X$ can be rewritten as a single exponential expression using the equation

$$A^X \bullet B^X = (A \bullet B)^X$$

You may want to ask students to explain this general principle in terms of repeated multiplication, in a manner similar to that used for the specific examples. A display like the one here illustrates how to rearrange and regroup the factors. Post this principle, with its explanation.

Question 2

The goal in Question 2 is to bring out that an expression of the form $A^X \bullet A^X$ can be simplified using either the additive law of exponents or the principle from Question 1 along with the law of repeated exponentiation explored in *Many Meals for Alice*. The two methods give equivalent results.

According to the additive law of exponents, $A^X \bullet A^X$ is equal to A^{X+X}, which equals A^{2X}. By the principle from Question 1, $A^X \bullet A^X$ is equal to $(A \bullet A)^X$, which equals $(A^2)^X$. And by the law of repeated exponentiation from *Many Meals for Alice*, $(A^2)^X$ is equal to A^{2X}. Thus the two approaches—the additive law of exponents and the principle from Question 1—lead to the same answer.

Question 3

Let students share their results on Question 3. It turns out that the only *positive integer* solutions to the equation $A^X = A \cdot X$ are

- $X = 1$ and $A =$ any number (because $A^1 = 1 \cdot A$, or $A = A$)
- $X =$ any positive number and $A = 0$
- $X = 2$ and $A = 2$ (the two solutions of the equations $Y = 2^X$ and $Y = 2X$; refer to the graph)

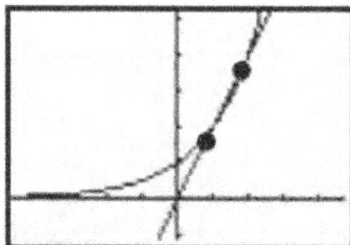

If X is a value other than 1, there will be a unique value of A that fits the equation, but unless $X = 2$, that value will be irrational. For instance, if $X = 3$, the value of A that fits the equation is $\sqrt{3}$.

What About 0^0?

The case for which X and A are both 0 presents a special problem. Bring out that in trying to define 0^0, there is a contradiction between two principles.

- On one hand, any power of 0 ought to be equal to 0. (You may need to look at examples such as 0^2 and 0^3 to clarify this principle.)
- On the other hand, we've defined expressions with an exponent of 0 to be equal to 1.

Because of this contradiction, the expression 0^0 is generally considered undefined. Calculators give an error message if you try to calculate 0^0.

Ask students, **Do you know of any other situations in which operations are undefined?** If no one thinks of any, remind them of division by zero and discuss why expressions like $5 \div 0$ are undefined. (The issue of undefined expressions will arise again when negative exponents are considered.)

Key Questions

In what ways can we write $3^7 \cdot 5^7$ as repeated multiplication?

How might the fact that there are the same number of 3s as 5s be useful?

Do you know of any other situations in which operations are undefined?

Having Your Cake and Drinking Too

Intent

Using the Alice metaphor and the rules they have derived so far, students discover why it makes sense to define expressions using negative exponents (such as 2^{-3}) as representing fractions.

Mathematics

For the sake of consistency, we define 2^0 as 1 and 2^{-B} as $\dfrac{1}{2^B}$. With these definitions, the rule $A^X \bullet A^Y = A^{X+Y}$ works even when $Y < 0$ and $|Y| \geq X$. In addition, the pattern that begins $2^3 = 8$, $2^2 = 4$, $2^1 = 2$ can continue, with $2^0 = 1$ and $2^{-1} = \dfrac{1}{2}$.

Based on their experience in this activity, students establish the principle that

$$2^c = \left(\frac{1}{2}\right)^B = 2^{C-B}$$ and use this principle to establish the definition $2^{-B} = \dfrac{1}{2^B}$.

Progression

Working in groups, students examine what Alice's height is multiplied by if she consumes both cake and beverage. Then they experiment with what happens if she consumes more beverage than cake. This results in a need to consider negative exponents, which are introduced during the discussion of the activity.

Approximate Time

60 minutes

Classroom Organization

Groups, followed by whole-class discussion

Doing the Activity

Introduce this task by posing the first question aloud to the class or as a question for groups to work on for a short period of time. What is Alice's height multiplied by if she consumes the same number of ounces of cake and beverage? Discuss students' ideas and record the equation they develop. Then have groups work on the remaining questions.

For Question 4, if a group comes up with the expression

$$2^c = \left(\frac{1}{2}\right)^B$$

encourage the members to look at their examples from Questions 2 and 3 to find a way to rewrite their answer as a power of 2 with a single exponent.

Discussing and Debriefing the Activity

You can begin the discussion after all groups have at least started Question 4. If some groups don't get to Question 5, you can deal with the issue it raises in the whole-class discussion.

If you haven't already talked about Question 1 as a class, begin by getting students to articulate the idea that equal amounts of cake and beverage "cancel out." They may be able to write a general equation to explain this cancellation, such as

$$2^N \bullet \left(\frac{1}{2}\right)^N = 1$$

Questions 2 and 3

For Question 2, students should determine that any combination in which the number of ounces of cake is 3 more than the number of ounces of beverage will work. They should find similar results for the two parts of Question 3.

Through their work on Questions 2 and 3, which start with a desired result and ask for combinations that yield that result, students will probably also recognize how to go in the other direction. Ask them to articulate the arithmetic they do when presented with a combination of cake and beverage. **How do you get from the number of ounces of cake and beverage to the effect on Alice's height?** They might say, for example, "Subtract the number of ounces of beverage from the number of ounces of cake, and take 2 to that power."

Question 4

Ask students to report their observations about individual cases as a general expression. **If C is the number of ounces of cake and B is the number of ounces of beverage, what happens to Alice's height?** They should be able to state that Alice's height is multiplied by 2^{C-B}.

Different groups may offer equivalent expressions that you can show to be equal. Or you may want to ask for an equation that shows the separate effects of the cake and the beverage in a single expression, such as

$$2^c = \left(\frac{1}{2}\right)^B = 2^{C-B}$$

Before going on, post the expression 2^{C-B} for combining cake and beverage. Students will refer to it as the basis for defining negative exponents.

Build confidence in the expression 2^{C-B} to help motivate the definition of negative exponents, as well as to reinforce earlier work with zero as an exponent. Ask, **Does the expression 2^{C-B} work if B and C are equal?** You might offer a specific example, such as Alice eating 4 ounces of cake and washing it down with 4 ounces of beverage.

Students should recognize that if $B = C$, Alice's height doesn't change; in other words, it is multiplied by 1. Because the expression gives 2^{4-4} and we have defined 2^0 as equal to 1, the expression does work when B is equal to C.

Question 5: The Case of Negative Exponents

To introduce the idea of negative exponents, ask, **What does the expression 2^{C-B} say in cases in which B is greater than C?**

Look at some specific cases. For example, **What does the formula say about the case in which Alice eats 0 ounces of cake and drinks 3 ounces of beverage?** There are two aspects to this question, and the key is bringing them together.

- Drinking 3 ounces of beverage multiplies Alice's height by $\left(\dfrac{1}{2}\right)^3$, or $\dfrac{1}{8}$.

- Substituting 0 for C and 3 for B in the expression 2^{C-B} tells us that Alice's height is multiplied by 2^{0-3}, which simplifies to 2^{-3}.

Once both aspects have been brought out, ask what this says about defining 2^{-3}. You may want to remind students that, like 2^0, the expression 2^{-3} cannot be defined in the usual way, in terms of repeated multiplication.

Students' work in developing the definition of 2^0 will probably lead them to conclude that it makes sense to define 2^{-3} as $\dfrac{1}{8}$. (*Rallods in Rednow Land* offers a different approach to the same conclusion.)

Present one or two more numeric examples, and then ask, **How can we generalize these results? In other words, how should we define 2^{-B}?** On one hand, this formula says to multiply Alice's height by 2^{0-B}, or 2^{-B}. On the other hand, drinking B ounces of beverage while eating no cake will multiply Alice's height by $\dfrac{1}{2^B}$.

Students should agree that, for the sake of consistency, it makes sense to define 2^{-B} as $\frac{1}{2^B}$. Emphasize that, as with the zero exponent, this is a definition. Students may prefer to write $\left(\frac{1}{2}\right)^B$ instead of $\frac{1}{2^B}$, which is fine.

Key Questions

What is Alice's height multiplied by if she consumes the same number of ounces of cake and beverage?

How do you get from the number of ounces of cake and beverage to the effect on Alice's height?

If C is the number of ounces of cake and B is the number of ounces of beverage, what happens to Alice's height?

Does the expression 2^{C-B} work if B and C are equal?

What does the expression 2^{C-B} say in cases in which B is greater than C?

What does the formula say about the case in which Alice eats 0 ounces of cake and drinks 3 ounces of beverage?

How can we generalize these results? In other words, how should we define 2^{-B}?

Rallods in Rednow Land

Intent

This activity will help students to appreciate the fact that quantities that grow exponentially grow very quickly.

Mathematics

The key mathematical idea in this task, in which students add the number of coins on each square of a chessboard, and the number of coins on each square is double the number on the previous square, is exponential growth. By the 31st square, the number of coins on that square alone has reached 2^{30}, or 1,073,741,824, which is more than the first choice for reward. The total number on the first 31 squares is $2^{31} - 1$, or 2,147,483,647. And there are still 33 squares to go.

To fully answer the question, students must devise a way to compute the sum of the geometric sequence

$$1 + 2 + 2^2 + 2^3 + \ldots + 2^n = \sum_{i=0}^{n} 2^i$$

Using patterns, they might realize that

$$1 + 2 = 3, \text{ or } 2^2 - 1$$
$$1 + 2 + 2^2 = 7, \text{ or } 2^3 - 1$$
$$1 + 2 + 2^2 + 2^3 = 15, \text{ or } 2^4 - 1$$

In general,

$$1 + 2 + 2^2 + 2^3 + \ldots + 2^n = 2^{n+1} - 1$$

Or, using the summation notation students have encountered before,

$$\sum_{i=0}^{n} 2^i = 2^{n+1} - 1$$

Progression

Working individually, students examine a classic problem about exponential growth by considering intuitively how the sum of a geometric series might compare to a very big number (1 billion). In the follow-up discussion, students share ideas about how to compute a geometric sum. (It is not intended that they develop a formal method.)

Approximate Time

5 minutes for introduction
20 minutes for activity (at home or in class)
15 minutes for discussion

Classroom Organization

Individuals, then groups, followed by whole-class discussion

Doing the Activity

Students are likely to encounter scientific notation on their calculators in their work on this activity. Some may have encountered this notation before; others may be unfamiliar with it. If they raise questions about the "strange" results on their calculators, offer a brief explanation and reassure them that they will learn more about this notation later in the unit.

Tell students you wish to share an interesting mathematical problem—related to exponents—that can be traced far back in history, and then describe the situation and the adviser's two choices. Ask students to share their initial, intuitive guesses as to which would be the better choice.

Discussing and Debriefing the Activity

Have students gather in their groups to compare ideas on Question 2. They do not need to find the sum of the sequence $1 + 2 + 4 + \cdots + 2^{63}$ to answer the question, because the 31st square by itself already has more than a billion rallods.

You might ask students to provide an expression for the number of rallods on the Nth square to clarify that it's 2^{N-1} and not 2^N.

Have a volunteer or two present their ideas on Question 3. One approach is to keep adding terms until the sum reaches 1 billion. Another is to recognize that each term is 1 more than the sum of the previous terms and then look for a term that is over half a billion. A third approach is to add the two or three largest terms and assume that this estimate is close enough. All three approaches show that 30 squares give just over a billion rallods.

Supplemental Activity

More About Rallods (extension) asks students to find a general formula for the sum of the first n powers of 2 and then has them explore other geometric sequences.

Continuing the Pattern

Intent

Students use a pattern approach to understand the definition of negative exponents. The activity sets the stage for a summative lesson on several recent topics.

Mathematics

Having explored the definition of negative exponents in several ways, students will now approach the topic by inspecting the pattern of decreasing exponents on a constant base. They will also reason through the equivalence of negative exponents on fractional bases to the positive (opposite) exponents on the corresponding reciprocal base. For example,

$$\left(\frac{1}{2}\right)^4 = \frac{1}{16} = 2^{-4}$$

In this context, students simplify complex fractions. They also consider calculator usage with regards to zero and negative exponents, and they return to the question of zero as a base.

Progression

Students work individually, and then as a whole class, to confirm that the pattern of exponential expressions agrees with the definition of exponentiation for negative exponents derived in *Having Your Cake and Drinking Too.* You then elaborate on the patterns, demonstrating ways to record equivalences involving complex fractions. The activity concludes with a discussion of calculators with regards to negative and zero exponents.

In their work in this unit, students are given several ways to approach a fairly abstract concept: that expressions with negative exponents are defined in terms of fractions. The activity *All Roads Lead to Rome* will review the various ways to think about extending the definition of exponential expressions beyond positive integer exponents.

Approximate Time

20 minutes for activity (at home or in class)
40 minutes for discussion

Classroom Organization

Individuals, followed by whole-class discussion

Doing the Activity

Tell students that in this activity they will discover another way of thinking about negative exponents.

Discussing and Debriefing the Activity

Begin the discussion by reviewing the sequence of results in Question 1. Ask students to describe the pattern, which can be articulated in various ways, such as, "Each value is half the one above it" or "Divide by 2 as you go down the list."

Encourage students to write the results involving negative exponents in several ways: as simplified fractions, as powers of $\frac{1}{2}$, and in the form $\frac{1}{2^B}$.

$$2^{-1} = \frac{1}{2}$$

$$2^{-2} = \frac{1}{4} = \left(\frac{1}{2}\right)^2 = \frac{1}{2^2}$$

$$2^{-3} = \frac{1}{8} = \left(\frac{1}{2}\right)^3$$

$$2^{-4} = \frac{1}{16} = \left(\frac{1}{2}\right)^4 = \frac{1}{2^4}$$

Ask them to relate the conclusions here to their previous work with negative exponents. **How do these results relate to your work on *Having Your Cake and Drinking Too?*** Use these examples to review and generalize the principle that we define 2^{-B} as $\frac{1}{2^B}$. Students should realize that the pattern just found for powers of 2 is another reason it makes sense to define expressions with negative exponents as we do. For many students, this is the most convincing and memorable way to think about negative exponents.

For Question 2, students should recognize that a similar pattern holds for any whole-number base and that, in general, it seems to make sense to define A^{-B} as equal to $\frac{1}{A^B}$.

For Question 3, they should realize that they have reversed the pattern from Question 1. Ask, **Does the general principle that A^{-B} is defined as $\frac{1}{A^B}$ apply** when A is $\frac{1}{2}$? Help students to understand that if this principle holds, we are saying, for example, that $\left(\frac{1}{2}\right)^{-4}$ should be defined as

$$\frac{1}{\left(\frac{1}{2}\right)^{-4}}$$

which suggests that $\left(\frac{1}{2}\right)^{-4}$ is equal to 16 (which is 2^4). This will require students to think about how to simplify a *complex fraction*—that is, a fraction in which the numerator or denominator is itself a fraction.

You may want to have students first look at the simpler case of

$$\frac{1}{\frac{1}{2}}$$

Here are two ways to think about simplifying this expression, both of which can be applied to any complex fraction of the form

$$\frac{\frac{a}{b}}{\frac{c}{d}}$$

- Multiply both the numerator (1) and the denominator $\left(\frac{1}{2}\right)$ by 2, giving

$$\frac{2 \bullet 1}{2 \bullet \frac{1}{2}}$$

which simplifies to $\frac{2}{1}$, which equals 2. Bring out that this is the same process students always use to create equivalent fractions, such as in expressing $\frac{1}{2}$ as $\frac{2}{4}$.

- Interpret the fraction as division, that is, as $1 \div \frac{1}{2}$, and use the "invert and multiply" rule to rewrite it as $1 \cdot \frac{2}{1}$.

Summing Up

Before using the additive law of exponents to confirm the definition, you may want to have a volunteer summarize the class's conclusions about defining expressions with negative integer exponents.

Students may initially give either a numeric example, such as $2^{-2} = \frac{1}{4}$, or a general principle, such as $A^{-B} = \frac{1}{A^B}$. Either way, help the class to state the general principle clearly, perhaps in this way:

An exponential expression with a negative exponent is defined by the equation $A^{-B} = \frac{1}{A^B}$.

Post this principle for reference.

Confirming the Definition Using the Additive Law of Exponents

Choose a particular expression with a negative exponent, such as 2^{-2}, and ask the class, **What is an example of the additive law of exponents that uses the expression 2^{-2}?** Ask for an example in which the exponent on the right side comes out positive, such as

$$2^{-2} \cdot 2^5 = 2^{-2+5}, \text{ or } 2^3$$

Have students verify that the new definition is consistent with the additive law of exponents by substituting the value of each expression: $\frac{1}{4}$ for 2^{-2}, 32 for 2^5, and 8 for 2^3. With these values substituted, the equation becomes the true equation

$$\frac{1}{4} \cdot 32 = 8$$

Then have students try an example or two in which all the exponents are negative, such as

$$2^{-3} \cdot 2^{-2} = 2^{-3+(-2)}, \text{ or } 2^{-5}$$

Substituting values based on the new definition gives the true equation

$$\frac{1}{8} \cdot \frac{1}{4} = \frac{1}{32}$$

A good special case to examine is one in which the exponents are opposites, such as $2^5 \cdot 2^{-5}$. Bring out that the exponents "cancel out" to give a sum of 0, while the exponential expressions themselves also cancel out to give a product of 1.

More Examples of Cake and Beverage

Review the general phenomenon of positive and negative exponents in terms of the Alice metaphor, perhaps by presenting several combinations of cakes and beverages—some with more cake, some with more beverage, some with equal amounts—and having students analyze the effect on Alice's height in two ways.

- Work with the two types of food sequentially (for example, first the cake, then the beverage) to determine the effect on Alice's height.
- Combine the cake and beverage into a single amount of one or the other, using the intuitive idea that "equal amounts of cake and beverage cancel each other." (Students may recognize that cake and beverage are analogous to hot and cold cubes, introduced in the Year 1 unit *Patterns.*)

Bring out that the two approaches give the same results. Help the class to verbalize the notion that in combining cake and beverage, they are treating the beverage as a kind of "negative cake." Therefore, the effect of B ounces of beverage, which is to multiply Alice's height by $\frac{1}{2^B}$, should be the same as the effect of $-B$ ounces of cake. According to the general formula students have developed, eating $-B$ ounces of cake should multiply Alice's height by 2^{-B}. In other words, $\frac{1}{2^B}$ and 2^{-B} should be equal.

Graphing y = 2ˣ

During the activity *Graphing Alice,* you posted a graph of points from the equation $y = 2^x$ for positive integer values of x. As students worked with zero as an exponent, they saw that defining 2^0 to be 1 seemed consistent with this graph. Now it's time to extend the graph to include negative integer values for x.

In that discussion, students speculated about what would happen if they extended the graph of $y = 2^x$ to include negative values of x. Ask volunteers to add new points to that graph, first for x with a value of 0 and then for negative integer values of x.

Although the scale of the earlier graph may make it hard to plot these points precisely, students should observe that the general shape of the first quadrant portion of the graph is consistent with these new points. As x moves from larger to smaller positive values, the y-values decrease in a way that fits smoothly with new points at $(0, 1)$, $\left(-1, \frac{1}{2}\right)$, $\left(-2, \frac{1}{4}\right)$, and so on.

Negative and Zero Exponents on Calculators

Ask students to verify that their calculators give results for zero and negative exponents that agree with the definitions developed so far.

For negative exponents, calculators will give results in decimal form. Students will need to verify that these are equal to the common-fraction values they have used in the definition. For example, if they evaluate 2^{-3}, they will get 0.125, which is equal to $\frac{1}{8}$.

Zero as a Base

Ask, **What happens if you use 0 as the base with a negative exponent? Why?** Students should get an error message. If time permits, this can lead to a good discussion of the issue of division by zero and why division by zero is undefined. At the same time, you can review the situation of 0^0 described in *In Search of the Law*.

Key Questions

How do these results relate to your work on *Having Your Cake and Drinking Too*?

Does the general principle that A^{-B} is defined as $\frac{1}{A^B}$ apply when A is $\frac{1}{2}$?

What is an example of the additive law of exponents that uses the expression 2^{-2}?

What happens if you use 0 as the base with a negative exponent? Why?

Negative Reflections

Intent

Students carefully review and summarize what they know about negative exponents in preparation for the upcoming topic of fractional exponents.

Mathematics

The work in *Extending Exponentiation* has extended the possible values for exponents from positive whole numbers—which can be interpreted as repeated addition—to zero and negative integers. This work has relied on the Alice metaphor, patterns, and the systematic development of rules using prior rules. In this activity, students pause to look back over and summarize this work.

Progression

Working individually, students summarize their work with integer exponents by writing an explanation for the definition along with some examples. They share their explanations with an adult, and a brief class discussion of these conversations concludes the activity.

Approximate Time

20 minutes for activity (at home)
5 minutes for discussion

Classroom Organization

Individuals, followed by whole-class discussion

Doing the Activity

Remind students that the past few activities have extended the definition for exponents from simple repeated multiplication for positive exponents ($2^5 = 2 \cdot 2 \cdot 2 \cdot 2 \cdot 2$) to also make sense for zero and negative values.

Discussing and Debriefing the Activity

You may want to invite a few students to briefly share how the adults they spoke with reacted to the ideas.

This is a good activity for assessing students' understanding. You will probably benefit from reading at least a selection of students' explanations, even if you choose not to grade the assignment. Students' explanations in Question 1 in particular will give you a good idea of how well they have understood the principles involved in extending the definition of exponentiation.

Curiouser and Curiouser!

Intent

Through these activities, students extend their growing understanding of exponents to include rational exponents.

Mathematics

In an approach similar to the development of meaning for zero and negative integer exponents, students' understanding of rational exponents will not be based on the idea that an exponent indicates repeated multiplication. Instead, students will use the Alice metaphor, numeric patterns, rules they developed in previous activities, and graphs of exponential functions to develop an understanding of this rule:

For all integer values x and y ($y \neq 0$), and for $b > 0$,

$$b^{x/y} = \sqrt[y]{b^x} = \left(\sqrt[y]{b}\right)^x$$

Progression

In *Curiouser and Curiouser!,* students first consider exponents that are unit fractions and then work with the more general case of any fraction. They then review all their rules for exponents and do an activity that sets up the study of logarithms in *Turning Exponents Around.* In addition, students begin work on the second POW of the unit.

A Half Ounce of Cake

It's in the Graph

POW 13: A Digital Proof

Stranger Pieces of Cake

Confusion Reigns

All Roads Lead to Rome

Measuring Meals for Alice

A Half Ounce of Cake

Intent

Students use the Alice metaphor to think about the meaning of fractional exponents. The discussion of the activity also provides some standard terminology related to exponents and checks that students are using their calculators appropriately.

Mathematics

In this activity, students learn the meaning of fractional exponents by equating them to what they already know about roots and to the additive law of exponents.

$$2^{1/2} = \sqrt{2} \text{ because } 2^{1/2} \cdot 2^{1/2} = 2^{1/2 + 1/2} = 2$$

The class discusses the language and notation of roots (which may be a review for some students), including the idea that the notation represents an exact value, whereas what a calculator reports is most often a rounded value.

This activity deals only with *unit fractions*—fractions with a numerator of 1 and a denominator that is a positive integer. More general fractional exponents will be explored in *Stranger Pieces of Cake.*

Progression

Students work on the activity in small groups and share their ideas with the whole class. A teacher-led discussion then provides conventions for exponent notation, connections to ideas about roots, and a review of working with roots and fractional exponents on the calculator.

Approximate Time

50 minutes

Classroom Organization

Groups, followed by whole-class discussion

Doing the Activity

To introduce the activity, ask, **What do you think would happen to Alice if she ate cake, but less than a whole ounce?** Students should quickly agree that she will grow, but to something less than double her original height.

Exactly how much would she grow? Say, for example, she eats exactly one half ounce. By what factor would her height be multiplied? One likely suggestion will be a growth factor of 1.5.

Let's say that piece was so good, she wants the rest of the ounce of cake. By what factor will she grow when she eats the second half ounce? Help students to recognize that if Alice eats another half ounce, her height will again be multiplied by the same factor. Encourage discussion so that students invest themselves in establishing this principle. Then put groups to work on the activity. *In your groups, design a way to check whether our initial guesses are correct.*

As they begin work, remind them to think about what Alice's height is *multiplied by* if she eats half an ounce of cake.

As soon as students realize that 1.5 doesn't work, many will begin searching for an appropriate multiplying factor by testing different numbers on the calculator.

One approach is to ask them to create an example of the **additive law of exponents** that shows Alice eating half an ounce and then another half an ounce, as suggested in Question 1. If someone suggests that this situation can be represented by the equation $2^{1/2} \cdot 2^{1/2} = 2^1$, suggest that students replace each instance of $2^{1/2}$ with a box and replace 2^1 with 2, and ask what number should go in each box (the same number in both) to make the equation true.

$$\square \bullet \square = 2$$

As needed, help students realize that they are looking for a number that if multiplied by itself gives a result of 2—that is, the square root of 2. They should be able to identify the number being described as $\sqrt{2}$.

You may want to ask the first groups that finish to prepare presentations.

Discussing and Debriefing the Activity

If a group has prepared to present Question 1, ask a representative to make that presentation. Emphasize that the presenter should talk about their group's exploration as well as their solution, rather than simply showing the number.

Remind the class of the general formula—that eating C ounces of cake multiplies Alice's height by 2^C—and ask how that formula applies to this situation. Be sure students recognize that the answer to Question 1 reveals how they should define $2^{1/2}$.

If no one suggests using the additive law of exponents, ask for an explanation defining $2^{1/2}$ based on that principle. Although students may have found that eating half an ounce of cake multiplies Alice's height by about 1.4, they may not make a connection between this number and the use of a fractional exponent.

Students will probably recall the term *square root.* Review the notation $\sqrt{2}$ if needed. They should be able to estimate $\sqrt{2}$, both by using the square-root key on a calculator and by guess-and-check, and thus realize that $2^{\frac{1}{2}}$ equals $\sqrt{2}$.

Question 2

After students have grasped that for half an ounce of cake they need a number whose square is 2, they should easily extend the idea to other fractional pieces of cake. Thus they should realize that if Alice eats a piece of cake that weighs a third of an ounce, her height will be multiplied by the number whose *third* power is 2. They can then use guess-and-check on their calculators to find a solution to the equation $x^2 = 2$.

The Language and Notation of Roots

Point out that the solution to $x^2 = 2$ is called the *square root of 2.* The solution to $x^3 = 2$ is called the *cube root of 2.* Introduce the notation $\sqrt[3]{2}$ for this number.

Follow up by introducing the notation $\sqrt[5]{2}$. Students should recognize that this is the solution to $x^5 = 2$. Use the phrase *fifth root of 2* for this number.

Point out that we could write $\sqrt[2]{2}$ for the square root of 2, but we don't. Also mention that the symbol $\sqrt{}$ is called the *radical sign.* (The word *radical* comes from a Latin word root that means "root.")

If students wonder why 1.41 is not sufficient for representing $\sqrt{2}$, explain that there is no decimal they can write whose square is exactly 2, so if they want to represent the number exactly, they need to use the symbol $\sqrt{2}$. (The issues of approximation and rounding are discussed in the unit *Do Bees Build It Best?,* especially in the activity *Falling Bridges.*)

Defining Fractional Exponents

You can now return to extending students' knowledge of fractional exponents. Review the general principle about Alice—that eating C ounces of cake multiplies her height by 2^C—and ask, What happens if Alice eats half an ounce of cake? Help students to understand that according to the general principle, Alice's height is multiplied by $2^{\frac{1}{2}}$, but their work in the activity shows that it should be multiplied by $\sqrt{2}$.

Remind students that as with negative and zero exponents, the "repeated multiplication" definition doesn't work for fractional exponents, so we need to *define* the expression $2^{\frac{1}{2}}$ by some other method. Students should recognize that their work suggests it makes sense to define $2^{1/2}$ as $\sqrt{2}$ and, similarly, $2^{1/3}$ as $\sqrt[3]{2}$ and $2^{1/5}$ as $\sqrt[5]{2}$.

You will probably want to get a summary of these ideas and post a general principle such as

$$A^{1/n} = \sqrt[n]{A}$$

Roots and Fractional Exponents on Calculators

Have students check that their calculators agree with the definition of $2^{1/2}$ as $\sqrt{2}$ by verifying that the calculators give the same answer for both.

Then have them explore how to find the value of roots and expressions using exponents that are unit fractions.

Some calculators use the symbol ^ (called a *caret*) for exponentiation, writing 2^5 for 2^5. It's often necessary to use parentheses around a fraction used as an exponent. For example, a calculator will probably interpret the expression 2^1/2 as $(2^1) \div 2$, which equals 1. The sure way to get the value of $2^{1/2}$ is to enter 2^(1/2).

Some scientific calculators have a $\boxed{\sqrt[x]{y}}$ key (often as a "second function" with the $\boxed{y^x}$ key). To represent $\sqrt[3]{4}$ using the $\boxed{\sqrt[x]{y}}$ key, you enter the *y*-value, press the $\boxed{\sqrt[x]{y}}$ key, and then enter the *x*-value. Thus a key sequence like

$$\boxed{4}\ \boxed{\text{2ndF}}\ \overset{\sqrt[x]{y}}{\boxed{y^x}}\ \boxed{3}$$

is used. (On some calculators, you would enter the *x*-value first.)

For calculators without such a key (as with many graphing calculators), one way to find roots other than square roots is by using fractional exponents. Another way is to use the MATH menu, with a key sequence like

$$\boxed{3}\ \boxed{\text{MATH}}\ \boxed{\sqrt[x]{\ }}\ \boxed{4}\ .$$

Help students make the connection between roots and fractional exponents. They should understand that the expression $a^{1/n}$ means the same thing as $\sqrt[n]{a}$.

Key Questions

What do you think would happen to Alice if she ate cake, but less than a whole ounce?

Exactly how much would she grow? If she eats exactly one half ounce, by what factor would her height be multiplied?

By what factor will she grow when she eats the second half ounce?

It's in the Graph

Intent

By using a graphical approach to make meaning of expressions like $2^{1/2}$, students reinforce their understanding of fractional exponents.

Mathematics

Students will compare graphs of the exponential function $y = 2^x$ and the linear function $y = x + 1$ to recognize from a visual perspective that, by the curved nature of $y = 2^x$, the value of y when $x = \dfrac{1}{2}$ must be less than 1.5. (Refer to the graph in "Discussing and Debriefing the Activity.") Students graph a few other exponential functions with similar questions in mind.

Progression

Students work on this activity individually, share ideas and questions in their small groups, and review their findings in a class discussion.

Approximate Time

20 minutes for activity (at home or in class)
15 minutes for discussion

Classroom Organization

Individuals, then groups, followed by whole-class discussion

Materials

It's in the Graph blackline master (transparency)

Doing the Activity

To introduce the activity, ask the class to restate what $2^{1/2}$ means in the Alice situation. Encourage at least two students to interpret the symbols in their own words.

Tell students that they will now consider what the graph of $y = 2^x$ suggests for the decimal value of $2^{1/2}$.

Discussing and Debriefing the Activity

The discussion of this activity will confirm the reasonableness of the definition that has been developed for an exponent of $\dfrac{1}{2}$. Students should realize that the estimates provided by their graphical analyses are consistent with this definition.

Begin with the comparison of graphs in Question 1, which will bring out graphically that $2^{1/2}$ is less than 1.5.

Students might also use a graphing calculator to graph the function $y = 2^x$ and use the Trace feature to get a more precise value for $2^{0.5}$ by finding the y-coordinate of the point for which x equals 0.5. The ZOOM menu will allow them to choose as much precision as they want. Make sure they realize that the y-coordinate has the numeric value of $\sqrt{2}$ and that their reading is an approximation of this.

Ask for a volunteer to present Question 2. He or she should note that the graph of $y = x + 1$ goes through the point (0.5, 1.5). Students have already observed that the graph of the equation $y = 2^x$ is curved, so when x is equal to 0.5, this graph is below the straight line through (0, 1) and (1, 2).

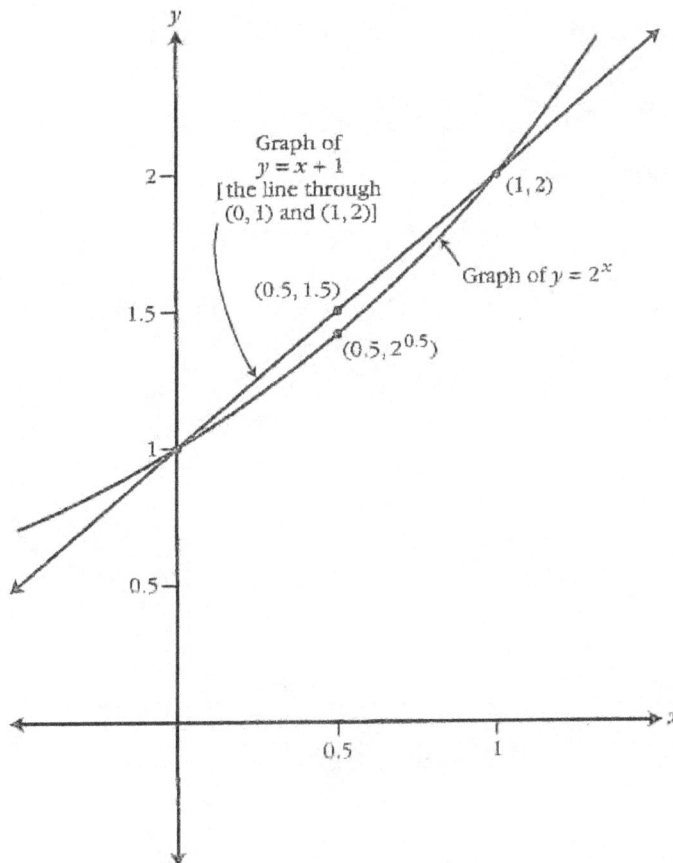

If time allows, use Question 3 to review the general graphs of other exponential functions. In particular, bring out that they all go through the point (0, 1). For part b, students should note that the graph of $y = 9^x$ seems to confirm that $9^{1/2} = 3$.

Finally, discuss the difference between the graph in part c and those in the other examples.

POW 13: A Digital Proof

Intent

To accomplish a proof for this problem, students must recognize key structures used to develop the puzzle.

Mathematics

The task for students is to determine all possible solutions to a number puzzle, which challenges them to think beyond simply guessing and checking until they find a solution that works. They complete the task by communicating an argument that their solution is unique. The deductive logic required is similar to that used in *POW 12: Logic from Lewis Carroll*.

Progression

This POW has two distinct phases of student work, followed by write-ups and presentations. Initially, students explore the puzzle, trying to find or narrow in on a solution. The second phase involves identifying structures of the puzzle that define the uniqueness of the solution. From these structures, students can write a proof.

Approximate Time

30 minutes for introduction
1–3 hours (at home)
20 minutes for presentations

Classroom Organization

Individuals and groups, follow by whole-class presentations

Doing the Activity

Read the activity as a class, and make sure students understand the directions. Have a volunteer explain the significance, for example, of putting a 3 in the box labeled 4. **What must happen if you put a 3 in Box 4?** Students should understand that this means there must be exactly three 4s used altogether in the boxes.

Have groups begin work on the problem. As they start to come up with possible solutions, encourage them to confirm aloud that each of the five digits they've placed in the boxes satisfies the requirements. Many will struggle to find a solution that works, failing to carefully consider the meaning of one or two numbers.

In their groups, students will begin to identify some of the constraints of the puzzle, which will help them develop a systematic process to narrow down possible solutions. Remind them that their main task is to prove they have found all possible solutions.

If groups find a solution, encourage them to identify what they learned about the puzzle that helped guide them toward that solution. These discoveries will likely be key parts of the proof they will write.

To help lead students to develop the argument they will need in the proof, ask, **Have you found all possible solutions? How do you know?**

The day before the POW is due, identify three students to prepare presentations.

Discussing and Debriefing the Activity

Have the three selected students make their presentations. Focus the presentations and discussion on the proof that the solution given is the *only* solution.

In one sense, the proof that the solution is unique consists of eliminating all other possibilities. However, because there are so many possible ways to fill the boxes (5^5, or 3,125), the other cases must be eliminated in an organized and systematic way.

This is an excellent problem for talking about proof. Students should recognize that saying "I couldn't find another solution" is not a proof that there are no others.

Key Questions

What must happen if you put a 3 in Box 4?

Have you found all possible solutions? How do you know?

Supplemental Activity

Ten Missing Digits (extension) expands the puzzle in this POW to ten missing digits.

Stranger Pieces of Cake

Intent

In this activity, students examine exponential expressions in which the exponent is not a unit fraction.

Mathematics

For all integer values x and y ($y \neq 0$), and for $b > 0$,

$$b^{x/y} = \sqrt[y]{b^x} = \left(\sqrt[y]{b}\right)^x$$

This rule makes sense given the other rules developed to this point in the unit. For example, using the law of repeated exponentiation,

$$b^{x/y} = \left(b^x\right)^{1/y} = \sqrt[y]{b^x} \text{ and } b^{x/y} = \left(b^{1/y}\right)^x = \left(\sqrt[y]{b}\right)^x$$

Progression

Working individually, students consider what an exponential statement like $2^{3/5}$ might mean, and then they generalize to a definition for fractional exponents. The follow-up discussion also explores negative fractional exponents and the idea that students have now articulated meaning for any rational number exponent for a positive base.

Approximate Time

5 minutes for introduction
20 minutes for activity (at home or in class)
35 minutes for discussion

Classroom Organization

Individuals, followed by whole-class discussion

Doing the Activity

Read through Question 1 of the activity with students to orient them to the task.

Many students will struggle with Question 1, especially if they don't recognize $\frac{3}{5}$ as "three one-fifths." If time allows, have them begin tackling the question in their groups prior to working on their own.

Discussing and Debriefing the Activity

Ask for volunteers to explain their work on Question 1. Students typically approach this by saying that eating a $\frac{3}{5}$-ounce piece of cake is the same as eating three $\frac{1}{5}$-ounce pieces. Because each $\frac{1}{5}$-ounce piece multiplies Alice's height by $\sqrt[5]{2}$, eating three such pieces will multiply her height by $\left(\sqrt[5]{2}\right)^3$. Thus it makes sense to define $2^{3/5}$ as $\left(\sqrt[5]{2}\right)^3$.

An alternate approach is to imagine Alice eating five pieces of cake, each weighing $\frac{3}{5}$ ounce. This is a total of 3 ounces of cake, and would thus multiply Alice's height by 2^3, or 8. Because five $\frac{3}{5}$-ounce pieces eaten together multiply her height by 8, one such piece would multiply her height by $\sqrt[5]{8}$.

The first approach interprets $2^{3/5}$ as $\left(\sqrt[5]{2}\right)^3$. The second approach interprets $2^{3/5}$ as $\left(\sqrt[5]{2}\right)^3$. Students can verify that these values are equal.

If no student suggests this idea, you might offer it yourself. Observing that $\sqrt[5]{2^3}$ should be equivalent to $\left(\sqrt[5]{2}\right)^3$ gives insight into the meaning of the notation.

One way to prove that $\left(\sqrt[5]{2}\right)^3$ is equal to $\sqrt[5]{2^3}$, based on the general laws of exponents, is to use the sequence of equalities

$$\left[\left(\sqrt[5]{2}\right)^3\right]^5 = \left(\sqrt[5]{2}\right)^{15} = \left[\left(\sqrt[5]{2}\right)^5\right]^3 = 2^3$$

which shows that $\left(\sqrt[5]{2}\right)^3$ is the fifth root of 2^3.

Question 2: Defining $2^{p/q}$

If students define $2^{3/5}$ as $\left(\sqrt[5]{2^3}\right)^3$, they will likely have little trouble generalizing to the idea that eating a piece of cake weighing $\frac{p}{q}$ ounces is like eating p pieces that each weigh $\frac{p}{q}$ ounces.

Because each $\frac{1}{q}$-ounce piece multiplies Alice's height by $\sqrt[q]{2}$, p such pieces should multiply her height by $\left(\sqrt[q]{2}\right)^p$. Therefore, it makes sense to define $2^{p/q}$ as $\left(\sqrt[q]{2}\right)^p$.

A student may ask, "What if q is 0? What is the 0th root of 2?" If so, point out that the exponential form would then be $2^{\frac{p}{0}}$, and the fraction $\frac{p}{0}$ is undefined. For similar reasons, the 0th root of 2, or of any number, is undefined.

Defining Negative Fractional Exponents

To extend the definition of exponentiation to include all rational numbers as exponents, ask whether anyone has an idea about this. **How should we define $2^{-1/2}$?** If needed, suggest there might be a way to use Alice's beverage for assistance. Remind students that drinking 3 ounces of beverage multiplies Alice's height by 2^{-3}, or $\left(\frac{1}{2}\right)^3$, and a whole ounce multiplies it by 2^{-1}, or $\frac{1}{2}$. Consequently, half an ounce should multiply her height by $2^{-1/2}$, or $\sqrt{\frac{1}{2}}$.

Does this definition fit the earlier principle that $A^{-B} = \dfrac{1}{A^B}$?

Work with students to help them recognize that

$$\sqrt{\frac{1}{2}} = \frac{1}{2^{1/2}}$$

In *Simply Square Roots* from the unit *Do Bees Build It Best?*, students worked with the general principle

$$\sqrt{\frac{a}{b}} = \frac{\sqrt{a}}{\sqrt{b}}$$

which shows that

$$\sqrt{\frac{1}{2}} = \frac{1}{\sqrt{2}}$$

The denominator $\sqrt{2}$ is equal to $2^{1/2}$, so

$$\sqrt{\frac{1}{2}} = \frac{1}{2^{1/2}}$$

The General Exponential Function

Another important idea that should come out of this discussion is that the function $y = 2^x$ makes sense for *any* number x. At this stage, students may have only a hazy notion that there are such things as irrational numbers, so you won't be able to discuss the "complete" exponential function, with a real number domain, in a formal sense. But they should realize that they have defined it for all rational exponents (for a positive base), and that should persuade them that they can make sense of any exponential expression with a positive base.

Bring out this point by asking, How could an expression like $2^{0.562}$ be interpreted in terms of the Alice situation? Students should be able to articulate that this number is what Alice's height would be multiplied by if she ate 0.562 ounce of cake. They should also recognize that theoretically they could find this number by thinking of this as 562 pieces, each weighing $\frac{1}{1000}$ ounce, and that each of those pieces would multiply Alice's height by $\sqrt[1000]{2}$, so that

$$2^{0.562} = \left(\sqrt[1000]{2}\right)^{562}$$

Similarly, $2^{-0.562}$ represents the factor by which Alice's height is multiplied if she drinks 0.562 ounce of beverage.

After exploring such examples, ask explicitly, Does the expression 2^x make sense for all values of x? Students should realize that it at least makes sense when x is a rational number. If they raise questions about irrational exponents, explain that the definition can be extended using repeated approximations and the concept of a *limit*.

The Complete Graph

This is a good occasion to look once again at the graph of $y = 2^x$ as a whole. Have students graph this function on their calculators and use the ZOOM menu and the Trace feature to check that the coordinates of points on the graph are consistent with the definitions they have formulated in extending the operation of exponentiation to include zero, negative, and fractional exponents.

The General Base and Exponent

Finally, bring out that the work just done with base 2 applies to any positive base. You might discuss a "random" example such as $0.416^{-6.78}$ to illustrate how such a general definition would work. Students can describe this as the factor by which Alice's height is multiplied if she drinks 6.78 ounces of base 0.416 beverage. Thus this number is equal to

$$\frac{1}{\left(\sqrt[100]{0.416}\right)^{678}}$$

Key Questions

How should we define $2^{-1/2}$?

Does this definition fit the earlier principle that $A^{-B} = \dfrac{1}{A^{B}}$?

How could an expression like $2^{-0.562}$ be interpreted in terms of the Alice situation?

Does the expression 2^{x} make sense for all values of *x*?

Supplemental Activities

Exponential Graphing (reinforcement) offers students more opportunities to examine the graphs of exponential functions.

Basic Exponential Questions (extension) raises challenging questions about inequalities involving exponential expressions. Question 2 is intentionally trivial (as it involves base 1). Question 3 follows up with a similar but more complicated problem. Question 4 is quite difficult to solve in general. The only whole-number solutions are the cases in which *X* is 2 and *Y* is 4, and vice versa. Students may find explanations for why there are no other solutions.

Confusion Reigns

Intent

Students review and explain the general principles for exponents developed during the unit. They come to the recognition that such laws eventually boil down to understanding what exponents are and that the laws can be re-created by writing out the exponential expressions using repeated multiplication.

Mathematics

Students review three laws of exponents, developed in the activities *Piece After Piece, Many Meals for Alice,* and *In Search of the Law,* with an emphasis on the justification for each law.

$$A^x \bullet A^y = A^{x+y}$$

$$A^x \bullet B^x = \left(A \bullet B\right)^x$$

$$\left(A^x\right)^y = A^{xy}$$

Progression

Students work individually to evaluate a number of proposed rules for working with exponents. In the follow-up discussion, they confirm their understanding and identify an associated law of exponents (or "nonlaw") for each question.

Approximate Time

20 minutes for activity (at home or in class)
20 minutes for discussion

Classroom Organization

Individuals, then groups, followed by whole-class discussion

Doing the Activity

Acknowledge that students have discovered many general laws about exponents that always hold. Tell them that in this activity, they will review some of those laws.

Discussing and Debriefing the Activity

You may want to let students spend some time in groups sharing ideas. They will be able to verify which equations are numerically correct by doing the arithmetic, so the focus of the discussion can be on explanations and finding general rules. Although the principles needed for Questions 2 and 3 were discussed earlier (in *Many Meals for Alice* and *In Search of the Law*), this is a good opportunity to review them.

The comments listed here summarize the key mathematical ideas to keep in mind during the discussion.

Question 1

There are no simple rules for adding numbers with exponents, whether the bases are the same or different. This "nonrule" may be worth remembering.

Question 2

The general principle for multiplying when the exponents are the same (whether the bases are the same or different) is

$$A^B \cdot C^B = (A \cdot C)^B$$

which is in line with Lara's idea. Have someone use repeated multiplication to explain in detail how Lara's example works. As discussed in connection with *In Search of the Law,* students should be able to write $2^3 \cdot 5^3$ as a product of individual factors, like this

$$2 \cdot 2 \cdot 2 \cdot 5 \cdot 5 \cdot 5$$

and then rearrange the factors in pairs, like this

$$(2 \cdot 5) \cdot (2 \cdot 5) \cdot (2 \cdot 5)$$

to show that this is equal to 10^3.

Question 3

Help students relate this question to their work on *Many Meals for Alice.* They should realize that Jen has the right idea by finding a general rule like

$$\left(A^B\right)^C = A^{BC}$$

Have them demonstrate how this works using a numeric example and repeated multiplication. For instance, $(7^2)^3$ means $7^2 \cdot 7^2 \cdot 7^2$, and each factor of 7^2 is equal to $7 \cdot 7$, so

$$(7^2)^3 = 7^2 \cdot 7^2 \cdot 7^2 = (7 \cdot 7) \cdot (7 \cdot 7) \cdot (7 \cdot 7) = 7^6$$

The "outside exponent" of 3 means there are three sets of 7s, and the "inside exponent" of 2 means there are two 7s in each set. Students should be able to explain that three sets of 7s with two 7s in each set is a total of six 7s, as $2 \cdot 3$ equals 6. And because the six 7s are multiplied together, the result is equal to 7^6.

So what's important here? It will be nice if students use either of these equations when they come across such situations:

$$A^B \bullet C^B = \left(A \bullet C \right)^B$$

$$\left(A^B \right)^C = A^{BC}$$

What's more important are the realizations that such laws of exponents eventually boil down to understanding what exponents are and that they can be re-created if one takes the time to write out the exponential expressions using repeated multiplication.

All Roads Lead to Rome

Intent

Students have used several approaches to gain an understanding of the extension of exponentiation beyond positive integer exponents. This activity gives them a chance to review and reflect on this variety of perspectives and will give you information on their ability to synthesize various approaches.

Mathematics

Students summarize how each of several approaches can be used to extend exponentiation to zero, negative, and fractional exponents. For example, to find $32^{2/5}$, they can use the rules derived in this unit to conclude

$$32^{2/5} = \left(32^{1/5}\right)^2 = \left(\sqrt[5]{32}\right)\left(\sqrt[5]{32}\right) = 2 \cdot 2 = 4$$

The unit provides a number of ways to make sense of this computation. According to the Alice metaphor, we want to know what to multiply Alice's height by if she eats $\frac{2}{5}$ ounce of base 32 cake. For 1 ounce, we would multiply by 32. For $\frac{1}{5}$ ounce, we would multiply by 2, as doing this 5 times would result in multiplying by 32. She will be eating $\frac{1}{5}$ ounce twice, so we should multiply by $2 \cdot 2$, or 4.

- Using the additive law of exponents, $32^{1/5}$ is the number we would multiply by itself 5 times to get 32, so $32^{1/5} = 2$. Then $32^{2/5} = 2^2 = 4$.

- Using the graph of $y = 32^x$, we can trace to find that y is 4 when x is $\frac{2}{5}$, or 0.4.

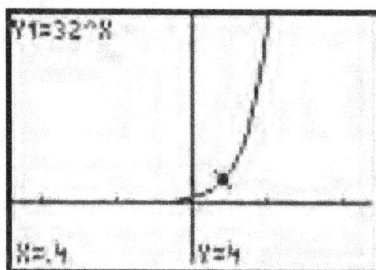

Progression

Students will explore this activity in small groups, with each student recording her or his own responses. A short conclusion organized around group presentations confirms their findings. Students' work in this activity will be included in their unit portfolios.

Approximate Time

50 minutes

Classroom Organization

Groups, followed by whole-class discussion

Doing the Activity

Tell students that this activity will be an opportunity to review and summarize their recent work with exponents. The reports they produce will be part of their unit portfolios.

Tell students that they do not have to explain every problem in Question 2 using all four methods. If they get bogged down with a particular explanation, they should move on and return to it later.

Discussing and Debriefing the Activity

Although this activity reviews the various approaches, it is not intended to lead to a detailed discussion of each problem. If time is limited, you might omit discussion of Questions 2c and 2d.

Ask for groups to present their ideas. You might begin with two presenting groups for the first problem or two and then reduce to one presentation as students feel comfortable.

Measuring Meals for Alice

Intent

In this final activity of *Curiouser and Curiouser!,* students look for "missing exponents" in the Alice setting. Their work sets the stage for introducing the concept of a logarithm.

Mathematics

If Alice is 1 foot tall, how much base 2 cake should she eat to grow to 10 feet? This question translates into the equation

$$1(2^x) = 10$$

The task is to find the exponent x that will produce a result of 10. In other words, given the function $y = 2^x$, find x for a given y.

This "undoing" activity sets students up to understand the concept of a **logarithm,** because if $y = 2^x$, then $x = \log_2 y$. That is, the inverse of the exponential function is the logarithmic function. At this point in the unit, students will guess-and-check using their calculators to approximate the solutions.

Progression

Working individually, students write an equation to represent each problem and solve for the unknown exponent using guess-and-check methods. In a class discussion, they share a variety of strategies for solving the problems.

Approximate Time

5 minutes for introduction
20 minutes for activity (at home or in class)
15 minutes for discussion

Classroom Organization

Individuals, followed by whole-class discussion

Doing the Activity

Introduce this activity by posing a question to remind students how to set up an exponential equation for these situations. **If Alice begins at 5 feet tall and wants to grow to 40 feet, how much base 2 cake will she need to eat?**

Even though students may be able to recognize quickly that 3 ounces of cake will do the trick, ask them to work in their groups to write an equation with the desired amount of cake as the unknown. Have them find a solution or show that their "obvious" solution works. Then explain that this next activity involves this same idea but in a more complicated form.

Discussing and Debriefing the Activity

You might ask students to share their ideas in groups and then report to the class, emphasizing their thought processes behind writing the equations and finding the solutions.

Students should find that the answer to Question 1 is 3.3 ounces. Before exploring the relationship between this and Question 2a, get an explanation of how the presenter derived the solution. He or she might recognize it as the solution to the equation $2^x = 10$. Presumably students will have found the numeric value by guess-and-check. (Logarithms will be introduced in *Sending Alice to the Moon.*) If the class had difficulty solving the problem, you might begin by asking for a rough approximation. Between which two whole numbers does the answer lie, and why?

The main focus of Question 2 should be on the connection between it and Question 1. Students should be able to explain, perhaps with some support from you, why the answer to Question 2 should be exactly twice that for Question 1.

Encourage a variety of explanations. Here are two possible approaches, one based on the Alice metaphor and one on laws of exponents.

- If eating 3.3 ounces of cake multiplies Alice's height by 10, eating this amount twice will multiply her height by 10 and then by 10 again. Therefore, eating 6.6 ounces multiplies her height by 100.
- The answer from Question 1 reveals that $2^{3.3}$ is approximately equal to 10, which means that $2^{6.6} = 2^{3.3 + 3.3} = 2^{3.3} \cdot 2^{3.3} \approx 10 \cdot 10 = 100$.

If students have given clear explanations for Questions 1 and 2, you may choose to discuss Questions 3 and 4 only briefly. In Question 3, Alice should drink about 1.6 ounces of beverage. In Question 4, she will become 3.8 feet tall.

Key Questions

If Alice begins at 5 feet tall and wants to grow to 40 feet, how much base 2 cake will she need to eat?

Between which two whole numbers does the answer lie, and why?

Supplemental Activity

Alice's Weights and Measures (extension) explores issues of approximation. When a measurement is only an approximation, what effect does it have on computations that make use of that measurement? This activity makes a good follow-up to *Measuring Meals for Alice.*

Turning Exponents Around

Intent

These final activities raise two important mathematical ideas related to exponents, as well as ask students to reflect over their work throughout the unit.

Mathematics

Students focus on two important mathematical ideas in these activities.

- The inverse of the exponential function $y = b^x$ is the logarithmic function $x = \log_b y$. The logarithmic function can be used to find the missing value of the exponent in an exponential equation. The graph of the logarithmic function is the reflection of the graph of the exponential function across the line $y = x$.
- Exponents can be used to write large and small numbers in **scientific notation,** as a number between 1 and 10 multiplied by an integer power of 10.

Progression

Students begin these activities by defining and exploring exponential functions. They then delve into scientific notation. They compile unit portfolios and take end-of-unit and end-of-year assessments. Finally, they participate in a class discussion of the big mathematical ideas in Year 2 of the curriculum.

Sending Alice to the Moon

Alice on a Log

Taking Logs to the Axes

Base 10 Alice

Warming Up to Scientific Notation

Big Numbers

An Exponential Portfolio

All About Alice Portfolio

Sending Alice to the Moon

Intent

Students search for "missing exponents" in the Alice setting, working particularly with base 10. The follow-up discussion leads to the development of the concept of a logarithm, to which students are introduced as a tool to solve for an unknown exponent. Students also identify some of the convenient logarithmic properties of working in base 10.

Mathematics

Drawing on their work in *Measuring Meals for Alice,* students begin this activity by solving a couple of exponential equations, in base 10 this time, which leads to the introduction of **logarithms.**

When confronted with equations of the form $10^x = 239{,}000$, students must find the power of 10 that results in the given value—in this case, 239,000. The answer is the base 10 logarithm of 239,000, which is written $\log_{10} 239{,}000$. And

$$\log_{10} 239{,}000 \approx 5.378 \text{ because } 10^{5.378} \approx 239{,}000$$

This activity includes an introduction to the terminology and notation of logarithms. It emphasizes the importance of base 10 logarithms and also brings out that numbers between 0 and 1 have negative logarithms.

Progression

Working individually, students are expected to solve these questions involving powers of 10 using guess-and-check on their calculators. The subsequent class discussion briefly resolves any questions and then moves into a mini-lecture introducing logarithms.

Approximate Time

20 minutes for activity (at home or in class)
40 minutes for discussion

Classroom Organization

Individuals, followed by whole-class discussion

Doing the Activity

This activity requires little or no introduction.

Discussing and Debriefing the Activity

To convey a sense of what the questions are asking, you might open the discussion by asking students between which two integers each answer lies. Save the answers to these questions for use in the discussion of logarithms that follows. In Question

1, Alice needs about 5.38 ounces of cake. In Question 2, she needs about 9.56 ounces of beverage.

Introducing Logarithms

Bring out that the questions in this activity and Questions 1 to 3 in *Measuring Meals for Alice* are quite similar. They all ask what power to raise a particular base to in order to get a certain result. Mention that this type of question gets asked a lot, in many different contexts, so people have devised special terminology and notation for it.

Explain that the solution to the equation $10^x = 239,000$, from Question 1, is represented by the expression

$$\log_{10} 239,000$$

Explain that this expression is read, "log, base 10, of 239,000" (or "log of 239,000 to base 10") and that *log* is an abbreviation for *logarithm*.

Ask, **What would this number mean to Alice?** Students should be able to articulate that it tells the amount of base 10 cake Alice needs to eat to multiply her height, which started as 1 mile, by 239,000.

Introduce a simpler example. **What does $\log_2 8$ mean to Alice? What does it mean as an exponential equation?** Students should realize that the expression represents how much base 2 cake Alice must eat to multiply her height by 8, as well as the solution to the exponential equation $2^x = 8$.

Using the Alice metaphor or the equation itself, students should figure that $\log_2 8 = 3$. Display this equation, and review how to read logarithm expressions.

Ask students to make up some logarithm expressions of their own and interpret them in terms of exponents. They should be able to articulate that an expression like $\log_a b$ represents "the power you raise a to in order to get b." That is, it represents the solution for x in the equation $a^x = b$.

What does the expression $\log_a b$ mean in Alice's situation? Students should recognize that a describes the type of cake (so we read the "a portion" of the expression as "base a") and b describes what is happening to Alice's height. The expression answers the question, "How much base a cake should Alice eat to multiply her height by b?"

Base 10 Logarithms

Tell students that because our number system is a base 10 system, logarithms for base 10 are widely used in mathematics and science, and most calculators have a key that gives base 10 logarithms.

Have students find some base 10 logarithms on their calculators, starting with verifying their answer to Question 1.

You might mention that logarithms to other whole-number bases are less commonly used and are in this unit only to provide more examples of logarithms. Point out that for whole-number bases other than 10, students cannot get the logarithm from a single key, so they should use guess-and-check, as they probably did for base 10 on the activity.

Estimating Base 10 Logarithms

To clarify the significance of using base 10 logarithms, you might ask, If the base 10 logarithm of a number is between 4 and 5, how big is the number? Students should conjecture that the number is between 10^4 and 10^5, or at least 10,000 and less than 100,000, which means it is a five-digit number. It may take several examples for them to make the connection between the size of the logarithm and the number of digits.

Then turn the process around and have students estimate the logarithm of a given whole number. Between which two whole numbers is the logarithm of 273,189? Students should realize that because this is a six-digit number, its logarithm must be between 5 and 6. As they develop this awareness, be sure they realize that this principle applies only to logarithms in base 10.

Work your way down to numbers between 1 and 10 as preparation for the discussion of logarithms for numbers between 0 and 1. Students should recognize that a number between 1 and 10 has a base 10 logarithm between 0 and 1.

Negative Logarithms

Build on the pattern students have just noted by asking about the base 10 logarithms for numbers between 0 and 1. What's the approximate value of $\log_{10} 0.25$? Emphasize that you only want to know between which two integers the answer lies.

Students may realize intuitively that the answer should be between –1 and 0. If not, you might return to the pattern established for large whole numbers and ask, We know that a number between 10 and 100 has a logarithm between 1 and 2 and that a number between 1 and 10 has a logarithm between 0 and 1. What would you expect for the logarithm of a number between 0.1 and 1? You should be able to establish that $\log_{10} 0.25$ is negative.

Ask students to find $\log_{10} 0.25$ more precisely on their calculators. If they do this using the base 10 logarithm key (and get approximately –0.6), ask, How could you confirm this result using exponents? The goal is to get them to recognize that the statement $\log_{10} 0.25 \approx -0.6$ is equivalent to the statement $10^{-0.6} \approx 0.25$.

The primary goal here is for students to discover that numbers between 0 and 1 have negative logarithms and that the smaller the number, the more negative the logarithm.

Logarithms of Negative Numbers?

Take this opportunity to bring out that only positive numbers have logarithms. You might ask about a specific example, such as, *What is* $\log_{10}$ *-2.7?* If students try to do this on a calculator, they should get an error message (unless the calculator is set to deal with complex numbers).

Why doesn't -2.7 have a base 10 logarithm? If necessary, ask students to write an appropriate exponential equation. They should recognize that they are trying to solve the equation $10^x = -2.7$ and that this equation has no solution, because exponential expressions with base 10 (or any other positive base) give only positive values.

Logarithms and the Graph of $y = 10^x$

You can clarify all of these ideas by referring to the graph of the equation $y = 10^x$. Bring out that y-values between 0 and 1 on this graph correspond to negative x-values and that there are no points on this graph with negative y-values. (The graph of the base 10 logarithm function is discussed in the activity *Taking Logs to the Axes*.)

The Alice Approach

Follow up the discussion of logarithms for numbers less than 1 by relating the issue of logarithms back to Alice's situation. If needed, review the general idea that asking for a logarithm is similar to asking how much cake Alice should eat.

Then take one of the examples the class has discussed, such as $\log_{10} 0.001$, and ask, *How much base 10 cake should Alice eat to multiply her height by 0.001?* Students may respond that eating cake will make her taller, not shorter. If so, use that response to review the idea that drinking the beverage is similar to eating "negative cake."

Based on this principle, students can find $\log_{10} 0.001$ by recognizing that Alice needs to drink 3 ounces of beverage to multiply her height by 0.001 and then interpreting this as the same as eating "–3 ounces" of cake. Thus it makes sense to define $\log_{10} 0.001$ as equal to –3.

Key Questions

What does $\log_2 8$ mean to Alice? What does it mean as an exponential equation?

What does the expression $\log_a b$ mean in Alice's situation?

If the base 10 logarithm of a number is between 4 and 5, how big is the number?

Between which two whole numbers is the logarithm of 273,189?

What's the approximate value of $\log_{10} 0.25$?

A number between 10 and 100 has a logarithm between 1 and 2, and a number between 1 and 10 has a logarithm between 0 and 1. What would you expect for the logarithm of a number between 0.1 and 1?

What is $\log_{10} -2.7$? Why doesn't -2.7 have a base 10 logarithm?

How much base 10 cake should Alice eat to multiply her height by 0.001?

Supplemental Activities

A Little Shakes a Lot (reinforcement) explores an interesting real-world use of logarithms: earthquakes.
Who's Buried in Grant's Tomb? (extension) offers another setting in which students can explore the relationship between exponents and logarithms.

Alice on a Log

Intent

Students explore logarithms in the context of Alice and her cake and beverage.

Mathematics

In the Alice metaphor, the equation $10^x = y$ is understood to mean, "If Alice eats x ounces of base 10 cake, her height will be multiplied by y." In this activity, the questions reinforce a similar interpretation of $\log_{10} y = x$: "x is the number of ounces of base 10 cake Alice needs to eat to multiply her height by y."

The focus of the activity is on the relationship between logarithmic and exponential equations. Students convert exponential situations into logarithmic equations and then estimate the solutions by guessing and testing with the associated exponential equations.

Progression

Students begin this activity in the classroom, with the opportunity to work with peers, and then complete it individually. They return to their groups to share ideas and ask questions before reviewing the concepts explored in the activity as a class.

Approximate Time

5 minutes for introduction
20 minutes for activity (at home or in class)
15 minutes for discussion

Classroom Organization

Individuals, then groups, followed by whole-class discussion

Doing the Activity

Encourage students to read through the activity and then begin working in their groups, possibly jumping right to Question 2. After they have a start on the activity, they can complete it on their own.

Discussing and Debriefing the Activity

Give groups a brief time to come to a consensus about the answers before having students report on each question.

Ask presenters to include appropriate exponential equations in their presentations. For example, the class should note that Question 2c is equivalent to asking for the solution to the equation $10^x = 50$, and the presenter should give the expression $\log_{10} 50$ to represent the answer.

For Questions 2e and 2f, students might use either 10 or $\dfrac{1}{10}$ as the base. For instance, they might interpret Question 2f in either of two ways:

- As asking for the solution to the equation $10^{-x} = \dfrac{1}{4}$. Then $-x$ is $\log_{1/10}\left(\dfrac{1}{4}\right)$, and x is $-x = \log_{1/10}\left(\dfrac{1}{4}\right)$.

- As asking for the solution to the equation $\left(\dfrac{1}{10}\right)^x = \dfrac{1}{4}$. Then $x = \log_{1/10}\left(\dfrac{1}{4}\right)$.

You might try to elicit both approaches, both of which give approximately 0.60 for x.

Finally, tell students that they will sometimes encounter just "log" rather than "$\log_{10}$." When no base is written, the assumption is that it is base 10.

Taking Logs to the Axes

Intent

Students sketch graphs of logarithmic functions and compare them with each other and with graphs of exponential functions. The graphs of both types of functions are not carefully examined in this unit, but students will become aware of some general properties.

Mathematics

The logarithmic and exponential functions are inverses, so their graphs are reflections across the line $y = x$. That is, if the point (a, b) is on the graph of $y = 2^x$, then the point (b, a) is on the graph of $y = \log_2 x$.

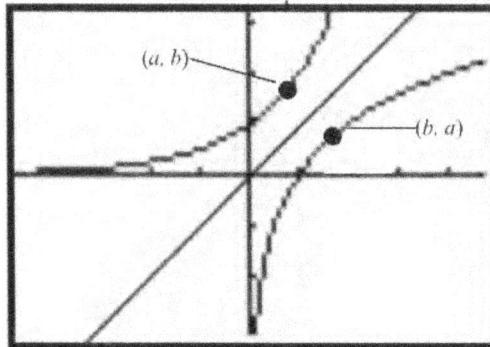

In this activity, students examine the graphs of logarithmic functions and their connections to those of exponential functions. In particular, they investigate how the graph changes as the base changes and recognize the symmetrical relationship between the graph of a logarithm function and that of the corresponding exponential function.

Progression

This activity opens with students guessing what the graph of a logarithmic function might look like. They then work individually or in small groups to explore such graphs, comparing graphs of different bases as well as graphs of related exponential and logarithmic functions.

Approximate Time

20 minutes for activity
10 minutes for discussion

Classroom Organization

Individuals or small groups, followed by whole-class discussion

Doing the Activity

This activity requires little or no introduction.

If students work on the activity in groups, bring the class together for discussion when most groups have done at least some work on Question 4.

As groups finish Questions 1a, 1b, and 2, you can choose a group to present the graph for each part, including discussing the scales.

Discussing and Debriefing the Activity

Have students briefly present the graphs for Questions 1a, 1b, and 2, or jump right to Questions 3 and 4. Encourage a variety of comparisons. For instance, they might point out that as the base for a logarithm function increases, the portion of the graph for values of x greater than 1 gets "flatter."

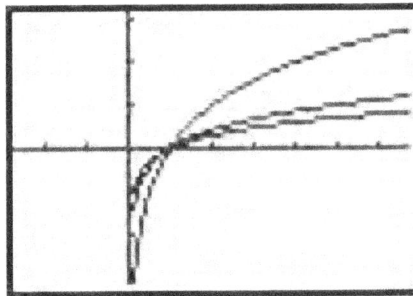

If no one points out the relationship between the graphs of a logarithm function and the corresponding exponential function (they are reflections across the line $y = x$), try to elicit this idea by having students identify specific points on each. For example, if they used the point (8, 3) for the graph of $y = \log_2 x$, bring out that this point fits this equation because $2^3 = 8$, which means (3, 8) is on the graph of the function $y = 2^x$. Using examples like this, help students realize that if (a, b) is on the graph of $y = \log_2 x$, then (b, a) is on the graph of $y = 2^x$ and the two graphs are symmetrical with respect to the line $y = x$. (Refer to the graph in the "Mathematics" section above.)

Base 10 Alice

Intent

This activity offers a series of base 10 problems in the Alice context designed to introduce students to standard ways of recording numbers in scientific notation.

Mathematics

Scientific notation is a standard method for writing very large or very small numbers by expressing them as a number between 1 and 10 multiplied by an integer power of 10. For example, the mass of the earth is estimated to be about $5.97 \cdot 10^{24}$ kilograms (that huge number starts with 597 and ends with 22 zeros), and today's computer processor transistors are about $4.5 \cdot 10^{-10}$ meter to $6.5 \cdot 10^{-10}$ meter across (less than a one billionth of a meter). In each case, the number between 1 and 10 expresses the significant digits of the number (the number's *precision*), and the power of ten expresses its **order of magnitude.**

In the context of this unit, the number between 1 and 10 can be thought of as Alice's initial height. The order of magnitude can be thought of as the number of ounces of cake she eats.

Progression

Students work on the activity individually, followed by some time in groups to share ideas. The follow-up discussion introduces scientific notation. The next activity, *Warming Up to Scientific Notation,* provides additional examples for students to explore.

Approximate Time

20 minutes for activity (at home or in class)
30 minutes for discussion

Classroom Organization

Individuals, then groups, followed by whole-class discussion

Doing the Activity

This activity requires little or no introduction.

Discussing and Debriefing the Activity

Give groups a few minutes to discuss their results and then bring the class together for a discussion.

Let volunteers explain their answers to the three parts of Question 1. For part c, it would be good to get both "50 trillion" and "5 with 13 zeros after it" as ways of describing the answer.

Then ask, **How could you write your answers to Question 1 using exponents?** For example, students should be able to write the answer to part c as $5 \cdot 10^{13}$. Point out that this is much easier to write and to work with than the value 50,000,000,000,000.

Have one or two students share their answers to Questions 2 and 3, with the goal of getting statements involving powers of 10 and "counting zeros." For example, a student might answer Question 2 by saying that the height Alice grows to must be 5 times a particular power of 10 or must be 5 followed by a number of zeros.

To solidify this basic idea, ask students to express a variety of numbers using powers of 10. **How would you write 23,000 as a whole number times a power of 10?** Solicit more than one answer, such as $23 \cdot 10^3$ and $230 \cdot 10^2$.

Connect these expressions with the Alice situation by asking, **What does each of these expressions mean in terms of an initial height for Alice and eating a whole number of ounces of base 10 cake?** For example, $23 \cdot 10^3$ could represent a starting height of 23 feet and eating 3 ounces of cake.

Point out that the different ways to write a given number as "whole number times a power of 10" also represent different starting heights and amounts of cake. This is also a good time to bring out that certain combinations involve initial heights that are not whole numbers. For example, help students to realize that 23,000 could also represent a starting height of 2.3 feet and eating 4 ounces of cake.

Scientific Notation

Use the discussion of this activity to introduce the idea of **scientific notation.** The introduction described here is based on the assumption that students have little or no previous exposure to the concept. Because you may have students who do know something about scientific notation, begin by asking if anyone is familiar with the term and have volunteers describe what they know. Build on that foundation to elicit any key concepts not introduced by the volunteers.

Tell students that it is useful in mathematics and science to have a standard way to write numbers using powers of 10. That standard way is like having Alice start at a height of at least 1 foot and less than 10 feet.

Ask, **How would you write 162,000 in this standard form?** They should be able to come up with the answer $1.62 \cdot 10^5$. Tell them that this is called scientific notation and that, for convenience, we will informally refer to the notation 162,000 as "ordinary" notation.

Similarly, the two components of a number written in scientific notation, such as 1.62 and 10^5 in the example, might be informally referred to as the "number part" and the "power-of-ten part."

Order of Magnitude

Tell students that in science and other disciplines, we often want a very rough idea of how big something is. The exponent in scientific notation helps with this approximation.

Two numbers that have the same exponent when written in scientific notation or that differ by a factor of less than 10 are said to have the same **order of magnitude.** For example, the population of California (about 38 million, as of 2008) has the same order of magnitude as the population of New York (about 19 million), because both numbers would be written in scientific notation as $a \cdot 10^7$, where a is a number between 1 and 10.

On the other hand, the population of New York has a different order of magnitude than the population of either the whole United States (about 300 million, or $3.0 \cdot 10^8$) or Nevada (about 2.4 million, or $2.4 \cdot 10^6$).

Questions 4 and 5

Return to Questions 4 and 5, and ensure that the numeric answers are clearly understood. For example, students should be able to explain why, if Alice starts out 5 feet tall and drinks 4 ounces of base 10 beverage, she ends up 0.0005 foot tall.

Ask students to articulate the mathematics involved in this rewriting. Focus their attention on the fact that moving the decimal point one place within a number changes the place value for each digit by a factor of 10 or $\frac{1}{10}$, depending on the direction of the move.

Writing Decimals Less than 1 Using Powers of 10

Ask students, How would you write the number 0.0005 using a power of 10? The context of Questions 4 and 5 may elicit the answer $5 \cdot \left(\frac{1}{10}\right)^4$. If so, acknowledge that this is equal to 0.0005, but state that you want students to use 10 instead of $\frac{1}{10}$ as the base. This may require some reminders about negative exponents and the idea that beverage is like "negative cake." Students should be able to express the result of starting at 5 feet and drinking 4 ounces of beverage with the expression $5 \cdot 10^{-4}$.

Go through a series of beverage problems, such as asking how Alice might end up 0.0035 foot tall. Students should realize that she could have started at 35 feet and drunk 4 ounces of beverage, or started at 3.5 feet and drunk 3 ounces of beverage, and so on.

Ask students how they think a height of 0.0035 foot should be written using scientific notation. They should recognize that $3.5 \cdot 10^{-3}$ is the standard form.

Key Questions

How could you write your answers to Question 1 using exponents?

How would you write 23,000 as a whole number times a power of 10?

What does each of these expressions mean in terms of an initial height for Alice and eating a whole number of ounces of base 10 cake?

How would you write 162,000 in this standard form?

How would you write the number 0.0005 using a power of 10?

Warming Up to Scientific Notation

Intent

Students practice writing and interpreting numbers in scientific notation and identify general principles for doing arithmetic with numbers in scientific notation.

Mathematics

Scientific notation is a way to write numbers—especially very large or very small numbers—as a number between 1 and 10 multiplied by an integer power of 10. When computing with numbers in this form, we take advantage of the rules for computation with exponents that students have been deriving in this unit. For example,

$$(3 \cdot 10^4) \cdot (2 \cdot 10^7) = 3 \cdot 2 \cdot 10^4 \cdot 10^7 = 6 \cdot 10^{4+7} = 6 \cdot 10^{11}$$

$$(9 \cdot 10^3) \div (3 \cdot 10^{-4}) = 9 \div 3 \cdot 10^3 \div 10^{-4} = 3 \cdot 10^{3-(-4)} = 3 \cdot 10^7$$

In this activity, students will derive these procedures.

Progression

Students work on their own to practice using scientific notation and then work individually or in small groups to devise methods for computing with numbers in scientific notation. The activity concludes with a whole-class discussion focusing on general principles about multiplying and dividing scientific numbers.

Approximate Time

10 minutes for introduction
20 minutes for activity (at home or in class)
10 minutes for discussion

Classroom Organization

Individuals, then groups, followed by whole-class discussion

Doing the Activity

Begin by having students do Questions 1 and 2 on their own. In Questions 3 and 4, they will begin to derive some general principles for working with scientific notation. Read the instructions to Question 3 aloud, and emphasize that students should do these problems without a calculator. Have students begin with part a, encouraging them to share their methods, and continue as time allows.

Ask for volunteers to share their techniques for simplifying the problem in part a. Students can complete the remainder of the activity on their own.

Discussing and Debriefing the Activity

Have students go over Questions 1 to 3 in their groups.

Ask for volunteers to share ideas on Question 4. Encourage them to provide examples and to talk about what they did to notice the patterns they are reporting. Students are likely to arrive at ideas like these.

- To multiply numbers in scientific notation, multiply the "number parts" and add the exponents of the "power-of-ten parts."
- To divide numbers in scientific notation, divide the "number parts" and subtract the exponents of the "power-of-ten parts."

Students may not realize that after applying these principles, they sometimes need to make an adjustment to standardize their answer. For example, following the first rule for Question 3b gives $35 \cdot 10^3$, but the correct scientific notation is $3.5 \cdot 10^4$. If necessary, ask, What do you need to do to adjust your answer to put it in scientific notation?

For Question 5, ask whether anyone has questions about how to work with scientific notation on the calculator. If there are questions, you might pair students up to work through the difficulties.

Key Question

What do you need to do to adjust your answer to put it in scientific notation?

Big Numbers

Intent

This final mathematical activity of the unit offers further practice with techniques for multiplying and dividing numbers in scientific notation and places scientific notation in several real-world contexts.

Mathematics

This activity draws together work with exponents and laws derived for exponents as students solve a set of practical problems. The mathematical focus is on estimation and **order of magnitude.**

Progression

After creating techniques for multiplying and dividing scientific numbers in *Warming Up to Scientific Notation,* students work in small groups to apply these techniques to some challenging problems. It is crucial not that they solve all the problems, but that they gain experience in working with and recording solutions to problems involving very big and very small numbers.

Approximate Time

40 minutes

Classroom Organization

Small groups, followed by whole-class discussion

Materials

To enhance the activity, you may want to obtain the video *Powers of 10* and related videos, which dramatically emphasize the magnitude of change related to powers of 10.

Doing the Activity

Have students begin working in their groups. If they are frustrated, you may want to do Question 1 as a class. If necessary, review the facts that a mile is 5280 feet (Question 3), a kilogram is 1000 grams (Question 5), and a meter is 1000 millimeters (Question 8).

Discussing and Debriefing the Activity

You might ask members of different groups to present each answer. Students may do these problems in a variety of ways, and it is worthwhile to elicit other approaches after the presentations, at least for a few of the problems.

Students might answer Question 1 by writing 30 as $3 \cdot 10^1$ and expressing the answer as the quotient $(3 \cdot 10^1) \div (5 \cdot 10^{-7})$. This gives $0.6 \cdot 10^8$, which can be rewritten in scientific notation as $6 \cdot 10^7$.

Alternatively, they might reason that because the computer does a computation in $5 \cdot 10^{-7}$ second, it can do 10^7 computations in 5 seconds. Because 30 seconds is 6 times as long as 5 seconds, the computer can do $6 \cdot 10^7$ computations in 30 seconds.

For Question 2, you might have students write an expression that represents the answer without actually doing any arithmetic. For example, you could express the number of seconds per year as

$$60 \cdot 60 \cdot 24 \cdot 365$$

and the number of gallons per year as

$$(60 \cdot 60 \cdot 24 \cdot 365) \div 76{,}000$$

You can then have students consider ways to estimate this product, such as rewriting it as approximately

$$(6 \cdot 10^1) \cdot (6 \cdot 10^1) \cdot (2 \cdot 10^1) \cdot (4 \cdot 10^2) \div (8 \cdot 10^4)$$

You might note that as 24 has been rounded down and 365 has been rounded up, there is some balancing out.

Students might then multiply $6 \cdot 6$ to get 36, approximate $36 \cdot 2$ as 70 and $70 \cdot 4$ as 300, and then approximate $300 \div 8$ as 40. The powers of 10 combine to give 10^1, for a final estimate of $40 \cdot 10^1$, which is 400. (A more exact answer is about 415 gallons. If it's a leap year, the answer is just over 416 gallons.)

Here are approximate answers to the remaining questions.

- Question 3: about $2.53 \cdot 10^8$ years (or about 253 million years)
- Question 4: about $29,500
- Question 5: about $5 \cdot 10^{25}$ atoms
- Question 6: about $3.3 \cdot 10^5$ or 330,000 earths
- Question 7: about $3.7 \cdot 10^{17}$ inches
- Question 8: about $4.7 \cdot 10^{15}$ grains of sand (or about 4.7 quadrillion grains of sand)

Supplemental Activity

Very Big and Very Small (extension) asks students to identify and investigate more situations involving very big and very small numbers in contexts they find intriguing.

An Exponential Portfolio

Intent

Students begin compiling their portfolios by recording and justifying all the laws about exponents derived during the unit.

Mathematics

In this summative activity, students should recall the following laws about exponents and justify why each is true.

$$A^x \cdot A^y = A^{x+y} \text{ for any rational numbers } x \text{ and } y$$

$$\left(A^x\right)^y = \left(A^y\right)^x = A^{xy} \text{ for any rational numbers } x \text{ and } y$$

$$A^0 = 1 \text{ if } A \neq 0$$

$$A^{-x} = \frac{1}{A^x} \text{ if } A \neq 0$$

$$b^{x/y} = \sqrt[y]{b^x} = \left(\sqrt[y]{b}\right)^x \text{ if } b > 0, x \text{ and } y \text{ are integers, and } y \neq 0$$

Progression

In one of the first steps toward creating their unit portfolios, students work on their own to review their notes in order to summarize what they have learned about one of the major mathematical topics of the unit: laws about exponents. Afterward, they can compare their lists in groups or with the whole class.

Approximate Time

30 minutes for activity (at home or in class)
10 minutes for discussion

Classroom Organization

Individuals, followed by small-group or whole-class discussion

Doing the Activity

This activity requires no introduction.

Discussing and Debriefing the Activity

Have students share their work in their groups or as a whole class. If done as a class, let volunteers offer general laws about exponents. As each law is suggested, have the rest of the class decide whether the statement is true. Solicit as many different explanations as possible.

All About Alice Portfolio

Intent

Students compile their unit portfolios and write their cover letters.

Mathematics

This activity helps students to review the main mathematical ideas of the unit: extending the operation of exponentiation, laws of exponents, graphing, logarithms, and scientific notation. In particular, students select work that demonstrates understanding of the operation of exponentiation, laws of exponents, and graphing.

Progression

Students begin by reviewing their work and notes for the unit in class and then complete the activity by writing a cover letter summarizing the mathematics of the unit and their personal growth during the unit and Year 2 of IMP.

Approximate Time

20 minutes for introduction
25 minutes for activity (at home)

Classroom Organization

Individuals

Doing the Activity

Have students read the instructions in the student book carefully. Review your expectations for their portfolios.

Discussing and Debriefing the Activity

You may want to have students share their portfolios in their groups, comparing what they wrote about in their cover letters and the activities they selected.

Blackline Master

It's in the Graph

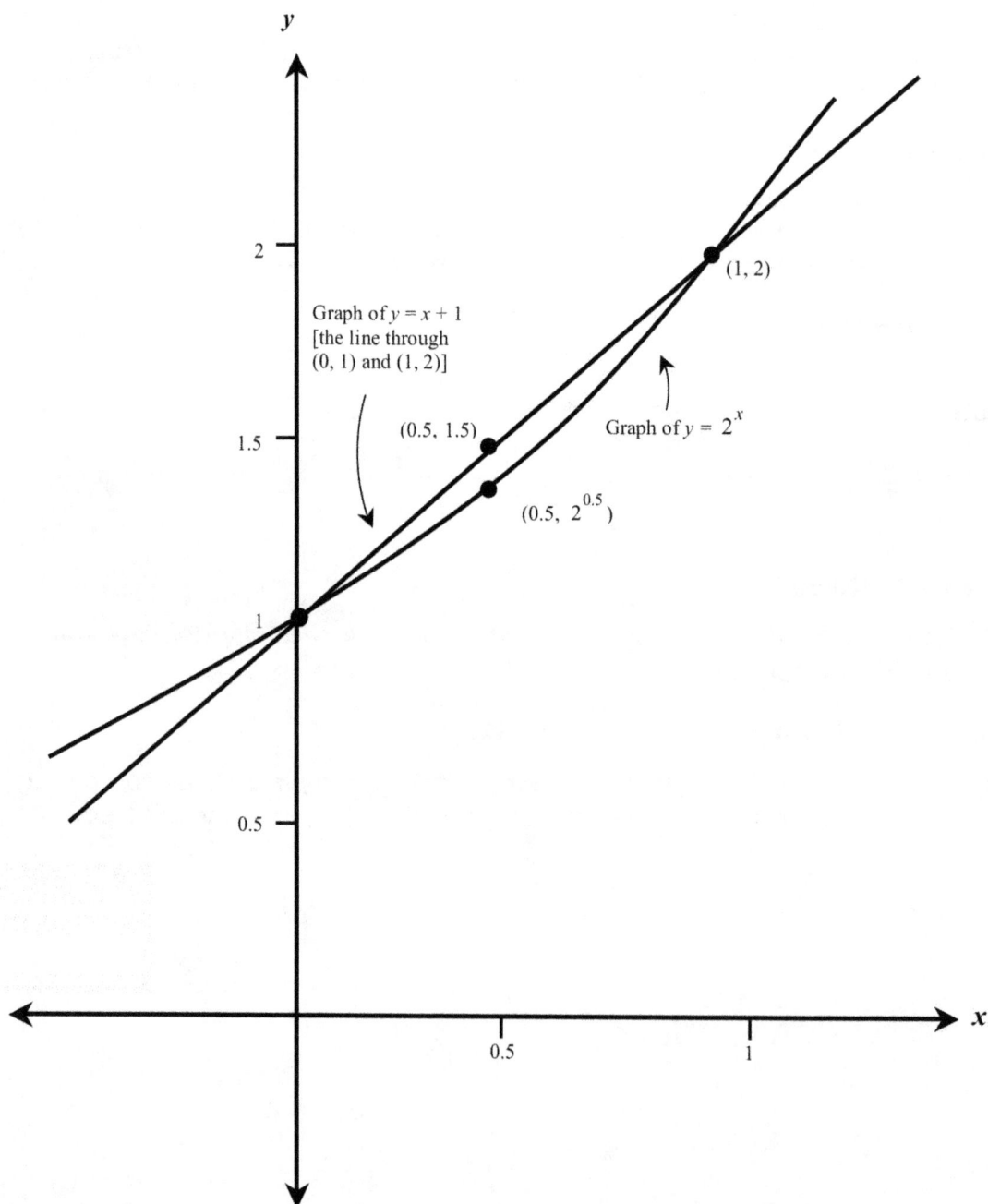

Graph of $y = x + 1$
[the line through
(0, 1) and (1, 2)]

(1, 2)

(0.5, 1.5)

Graph of $y = 2^x$

(0.5, $2^{0.5}$)

In-Class Assessment for *All About Alice*

Seven equations are given here. Some of the equations are true and some are false. Do parts a and b for each equation.

a.   State whether the equation is true or false.

b.   Explain your answer.

- If you think the equation is true, explain why. If possible, state and explain a general principle that the equation illustrates.
- If you think the equation is false, change the right side of the equation to make it true. Then explain why the new equation is true.

1. $10^5 \bullet 10^{12} = 10^{17}$

2. $\dfrac{3^8}{3^2} = 3^4$

3. $\sqrt{10^{16}} = 10^4$

4. $\left(5^2\right)^3 = 5^6$

5. $\dfrac{1.4 \bullet 10^8}{0.7 \bullet 10^4} = 2 \bullet 10^4$

6. $5^3 + 5^2 = 5^5$

7. $\log_2 8 = 3$

Take-Home Assessment for *All About Alice*

Part I: Graph It

1. Sketch the graph of the equation $y = 1.5^x$ from $x = -3$ to $x = 3$. Label at least five points with their coordinates.

2. Explain and show how to use your graph to estimate these values.

   a. $1.5^{-0.5}$

   b. $\log_{1.5} 2$

Part II: Far, Far Away

Give your answers to these questions in scientific notation, and explain clearly how you got them.

3. Light travels very fast, at approximately $1.86 \cdot 10^5$ miles per second. A *light-year* is the distance light travels in a year.

   a. About how many miles are there in a light-year?

   b. The star Betelgeuse is about $2.5 \cdot 10^{15}$ miles from the earth. About how many light-years is this distance?

4. An *astronomical unit* is the distance from the earth to the sun, which is approximately $9.3 \cdot 10^7$ miles.

   a. About how many astronomical units from the earth is Betelgeuse?

   b. Which is bigger: an astronomical unit or a light-year? About how many of one equals the other?

Part III: All Roads Lead to Understanding

The next two equations show how certain expressions with exponents are defined. For each equation, explain in as many different ways as you can why the expressions are defined the way they are. Give at least two explanations for each definition.

5. $4^0 = 1$

6. $5^{-3} = \dfrac{1}{125}$

All About Alice Guide for the TI-83/84 Family of Calculators

This guide gives suggestions for selected activities of the Year 2 unit *All About Alice.* The notes that you download contain specific calculator instructions that you might copy for your students. NOTE: If your students have the TI-Nspire handheld, they can attach the TI-84 Plus Keypad (from Texas Instruments) and use the calculator notes for the TI-83/84.

The primary calculator topics in *All About Alice* are exponents, logarithms, and scientific notation. As students begin to explore exponential functions, both the calculator's graphing feature and its ability to repeatedly execute the previous command will be very useful in illustrating the concepts involved.

Alice in Wonderland: The calculator's ability to recall and use a previous entry can provide another way to illustrate what happens as Alice eats several ounces of cake. Begin by entering a value for her initial height (1 meter is convenient) into the calculator and pressing ENTER. Now press ⊠ 2 ENTER to see the results of eating the first piece of cake. After the first ounce, every time you press ENTER, the calculator will display Alice's height after eating another ounce.

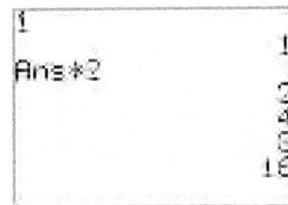

Be sure students learn how to use exponents both on the graphing calculator and on their personal calculators. On the graphing calculator, 7^2 can be entered simply as 7 x^2, and 7^3 is entered as 7 ∧ 3, whereas on many personal calculators, 7^3 is entered as 7 y^x 3.

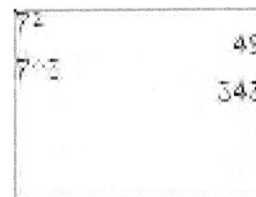

A Wonderland Lost: The calculator can help students determine the rule for Question 3. After you have reached the point in the discussion at which students recognize that multiplying the area by 0.9 is the same as subtracting 10 percent, you might use the next sequence of questions while illustrating on the overhead graphing calculator. Enter the initial area of 1,200,000 and press ENTER. Press ⊠ .9 ENTER and ask what this represents. Press ENTER several times, asking what the result represents after each time. Finally, ask what was done to 1,200,000 to get the last number that is displayed and how this could be written in a shorter form.

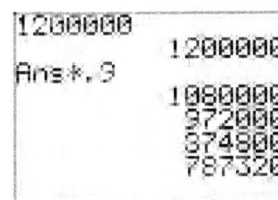

Here Goes Nothing: When discussing Question 4, let students enter **2^0** into their calculators to see that it really does give a result of 1. Some students will have done this right away, and you may need to remind them to explain why their answer makes sense in terms of the conclusions from the first three questions.

A New Kind of Cake: Question 5 is bound to stimulate some discussion of how the two functions $y = 2^x$ and $y = x^2$ compare for small values of x, because the curves cross each other a couple of times. Students will probably not have sketched their graphs in enough detail to see this clearly, and it is tricky to find a viewing window on the calculator that shows the two curves distinctly. These window settings work fairly well:

Xmin=1

Xmax=5

Ymin=0

Ymax=20

A heavier line style can be chosen for one of the functions, making it easier to differentiate between them. After this part of the curves has been discussed, change **Xmax** to **7** and **Ymax** to **100** to give a clear picture of how rapidly the two functions will diverge.

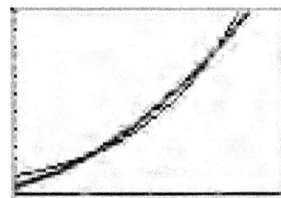

Piece After Piece: The effect of eating the cake in two stages can be illustrated quite clearly on the calculator during the discussion of this activity. Use a procedure similar to that described previously for *Alice in Wonderland,* but pause after 3 ounces and again after 5 more ounces. It will be very clear that the result would be the same without the pauses.

In Search of the Law: The discussion in the subsection "What about 0^0?" in the Teacher's Guide for this activity represents another opportunity to reinforce the vocabulary terms *range* and *domain.* Have students try 0^0 on the graphing calculator, and the calculator will display **ERR:DOMAIN.**

Rallods in Rednow Land: Students will probably encounter scientific notation on their calculators when working on this activity. Scientific notation will be covered later in this unit. At this time, students simply need to know that a display of, suppose, **1.844674407E19** means that the decimal belongs 19 places to the right of where it now is.

Keeping a running total of the number of coins on the board so far presents an interesting problem in Question 3. Writing down and reentering intermediate answers into the calculator is cumbersome and prone to error. Some of your students may enjoy the challenge of writing a program to answer Question 3 or to find the number of squares necessary to yield any given sum. The calculator entries listed here may also be used to do the calculation:

$1 \rightarrow x : 1 \rightarrow T$ ENTER   displays the total for the first square;

$X * 2 \rightarrow X : T + X \rightarrow T$ ENTER   displays the total for the first two squares. (*X* represents the number of coins on a square; *T* represents the total number of coins so far.)

Each time ENTER is pressed, the coins on one more square will be added. Simply count the number of times you press ENTER (including the first two times) until a sum greater than one billion is displayed.

Having Your Cake and Drinking Too: After the discussion of this activity, have students verify that the calculator gives the same result for 2^{-3} as it does for $\frac{1}{2^3}$. Remind them that when entering negative exponents, they must use the negative key, not the subtraction key.

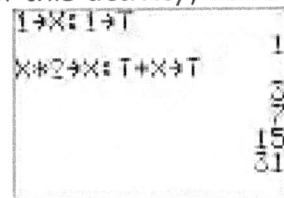

It's in the Graph: After students have decided what the graph of $y = 2^x$ should look like when it is extended for negative values of x, have them confirm their findings with a graph on the calculator.

There are two ways to set the calculator up to trace only to integer values of x. One way is with 8:**ZInteger** from the **Zoom** menu. This is used exactly like **Zoom In**, with the cursor and ENTER key used to define the center of the new screen. Another way is entering **1→ΔX** on the home screen and using an integer for **Xmin**. Press VARS, select the **Window** menu, and then choose **ΔX**. Have students trace to an x-value of zero and to negative values of x.

Ask students to return to the home screen by pressing 2ND [QUIT] and then to enter **0^−2**. They will get an error statement. Have them explain why they cannot use a negative power of zero.

Then ask what the graph of $y = 0^x$ will look like. Have students graph it on the calculator. They will seem to get a blank graph screen. The actual graph is covered by the axis. Turn the axes off by pressing 2ND [FORMAT], highlight **AxesOff**, and pressing ENTER.

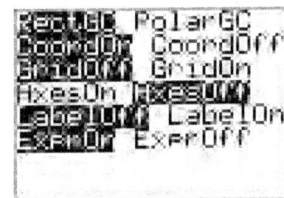

Press TRACE to return to the graph, which will appear as a horizontal ray along the positive x-axis. Tracing along the graph yields y-values of zero except where x is zero or negative. For $x \leq 0$, the calculator does not display a value for y, because it is undefined.

A Half Ounce of Cake: If students try to solve this activity by entering $2^{(1/2)}$ into the calculator, ask them to show another way of finding that answer. Explain that the object of this activity is not really to find the number by which Alice's height is multiplied, but to find it in a way that will lead to understanding what that fractional exponent really means.

Stranger Pieces of Cake: After the discussion of roots and fractional exponents, students need to become familiar with how to handle these on both the graphing calculator and their own calculators. Suggest that they use an expression whose value they know, such as $8^{1/3}$, to explore how each calculator works.

When entering a fractional exponent on the graphing calculator, students will have to be careful to use the parentheses. Otherwise, they will get the result of $(8^1)/3$ instead of $8^{(1/3)}$.

Many calculators have a $\sqrt[x]{y}$ function, often as a secondary operation of a y^x key. Students will need to experiment to determine in which order they need to enter the values for x and y on those calculators.

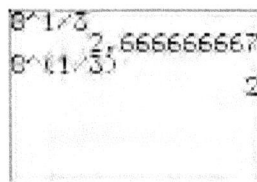

The graphing calculator has root functions under the **MATH** menu accessed by pressing MATH—there is a cube-root function and a $\sqrt[x]{}$ function.

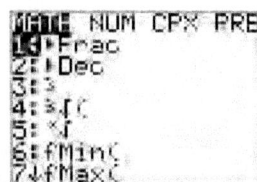

Have students compare the results of each of the following methods of finding the cube root of 8.

1.   Enter **8^(1/3)**.

2.   Press MATH, select **4:** $\sqrt[3]{}$ from the menu, and press 8 ENTER.

3.   Enter 3, press MATH, select **5:** $\sqrt[x]{}$ from the menu, and press 8 ENTER.

Measuring Meals for Alice: A nice alternative to guess-and-check on Question 1 is to graph the function $y = 2^x$ and trace to find the value of x that yields $y = 10$.

Alice on a Log: It will again be necessary for students to experiment with their own calculators to see how the logarithm function works and in which order the keys must be pressed. Have them use a logarithm for experimentation that they can easily interpret, such as $\log_{10} 1000$, which equals 3.

The graphing calculator has a LOG key with a secondary operation of 10^x. Enter the keystrokes in the same order in which you would write them. To find $\log_{10}1000$, press LOG 1 0 0 0 ENTER.

Inform students that when they write "log" without specifying a base, it is assumed to be base 10. The LOG key on the calculator will only find base 10 logarithms, so the base is never entered into the calculator. Base 10 logarithms and natural logarithms (which will be introduced in Year 3) are the only logarithms supported by the graphing calculator.

Taking Logs to the Axes: Question 2 asks students to use the graphing calculator to draw the graph of $y = \log_{10} x$. Choosing the viewing window appropriately will take some thought. Ask students to think carefully about what happens to y as x increases. Which will be larger, x or y? By how much? They will find that the range they select for x will have to be much greater than that for y in order to get a meaningful graph. Ask them to try to find a viewing window that shows large values of x as well as what happens when x is between 0 and 1 .

Question 4 of is difficult to illustrate on the graphing calculator—being limited to base 10 causes the graphs to be so steep that the curves appear to merge with the axes with many window settings. These settings below yield a window that allows a reasonable illustration for the functions $y = \log x$ and $y = 10^x$:

Xmin=−6

Xmax=8

Ymin=−2

Ymax=8

Warming Up to Scientific Notation: The Calculator Note "Scientific Notation on the Calculator" will introduce your students to how the graphing calculator handles scientific notation. They will still need to explore how scientific notation is handled on their personal calculators, but most are very similar.

Scientific Notation on the Calculator

You have probably already encountered scientific notation displayed on your calculator, but you may not have realized what it was. When a number is very large (or very small), the calculator automatically shifts to scientific notation. An easy way to note how it works is to enter the sequence of operations listed here: Enter the number 10 and press ENTER. Now multiply the result by 10 by pressing ✕ 1 0 ENTER. Now continue to multiply the result by 10 by simply pressing the ENTER key. Do this until the calculator shifts into scientific notation.

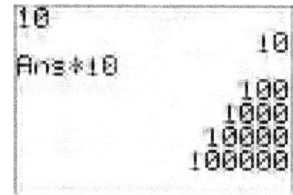

Notice that a display of **1E10** does not mean 1^{10} (which would only be 1), but $1 \cdot 10^{10}$ (which is 10 billion). (Similarly, **1E12** means $1 \cdot 10^{12}$. The base, 10, is fixed.) This notation with an **E** in the display is commonly used in calculators, but it is not conventionally used for results written on paper. When recording your answer from the calculator on paper, use the full scientific notation, which includes writing the base of 10 and its exponent.

You can enter numbers into the calculator in scientific notation using **EE** (enter exponent). It is the secondary operation of the comma key, which is located in the row above the 7 key. For example, enter $3.5 \cdot 10^{12}$ by pressing 3 . 5 2ND [EE] 1 2 ENTER.

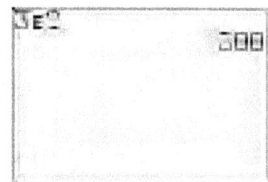

If you enter a smaller number in scientific notation, like $3 \cdot 10^2$, and then press ENTER, the calculator will switch back to standard notation to display the answer. You can force the calculator to display all answers in scientific notation by going to the MODE screen, highlighting **Sci,** and pressing ENTER. Try entering $3 \cdot 10^2$ again and note what happens.

If you have a personal calculator that is not a graphing calculator, experiment with scientific notation on that calculator as well. Pay special attention to the order of keystrokes necessary to enter a number in scientific notation with a negative exponent. Some calculators require that the exponent be entered first and then be made negative, while others require pressing the negative key before entering the number in the exponent. Some display the power of 10 without the **E**.

www.ingramcontent.com/pod-product-compliance
Lightning Source LLC
Chambersburg PA
CBHW051348200326
41521CB00014B/2515